AF411311

Corrosion in Concrete: Inhibitors and Coatings

Corrosion in Concrete: Inhibitors and Coatings

Editor

Luigi Coppola

MDPI • Basel • Beijing • Wuhan • Barcelona • Belgrade • Manchester • Tokyo • Cluj • Tianjin

Editor
Luigi Coppola
University of Bergamo
Italy

Editorial Office
MDPI
St. Alban-Anlage 66
4052 Basel, Switzerland

This is a reprint of articles from the Special Issue published online in the open access journal *Materials* (ISSN 1996-1944) (available at: http://www.mdpi.com).

For citation purposes, cite each article independently as indicated on the article page online and as indicated below:

LastName, A.A.; LastName, B.B.; LastName, C.C. Article Title. *Journal Name* **Year**, *Volume Number*, Page Range.

ISBN 978-3-0365-3301-8 (Hbk)
ISBN 978-3-0365-3302-5 (PDF)

Cover image courtesy of Luigi Coppola

Contents

About the Editor

Luigi Coppola, Civil Engineer, Associate Professor, University of Bergamo (Italy), Department of Engineering and Applied Sciences, and President of American Concrete Institute Italy Chapter. He has authored more than 335 original papers, and 21 books on admixtures for concrete, alternative low-carbon binders, waste management in concrete production, deterioration, durability and repair of concrete structures, mix design, deterioration and restoration of historical buildings, corrosion and protection of rebars in reinforced concrete structures. In June 2000, Luigi Coppola was awarded for "Outstanding and sustained contributions to enhance the Durability of Concrete" by the American Concrete Institute (ACI) and the Canadian Institute of Materials, Energy and Transportation (CANMET). He is the editor and member of numerous international journals and a chairperson of Thirteenth International Conference on Superplasticizers and other Chemical Admixtures in Concrete—Milan, Italy, June 5–8, 2022 and Fifteenth International Conference on Recent Advances in Concrete Technology and Sustainability Issues—Milan, Italy, June 8–10, 2022.

Editorial

Special Issue "Corrosion in Concrete: Inhibitors and Coatings"

Luigi Coppola [1,2,3]

[1] Department of Engineering and Applied Sciences, University of Bergamo, 24044 Dalmine, Italy; luigi.coppola@unibg.it; Tel.: +39-035-2052054
[2] UdR "Materials and Corrosion", Consorzio INSTM, 50121 Florence, Italy
[3] Consorzio Interuniversitario per lo Sviluppo dei Sistemi a Grande Interfase—Center for Colloid and Surface Science (CSGI), 50019 Sesto Fiorentino, Italy

Citation: Coppola, L. Special Issue "Corrosion in Concrete: Inhibitors and Coatings". *Materials* **2021**, *14*, 6211. https://doi.org/10.3390/ma14206211

Received: 28 September 2021
Accepted: 8 October 2021
Published: 19 October 2021

Publisher's Note: MDPI stays neutral with regard to jurisdictional claims in published maps and institutional affiliations.

The climatic changes that are taking place in recent years have increased awareness of the importance of environmental protection and the urgent need for industrial strategies aimed at a sustainable development. The construction industry has always been perceived as one of the sectors with the greatest environmental impact and therefore the cement and concrete production has developed several strategies to limit the environmental impact that largely derives from the binder production [1–3]. These strategies are aimed not only at reducing the environmental impact (carbon dioxide emissions, energy consumption and natural raw materials depletion) but also at improving the performance and durability of building materials.

Penetration of aggressive external agents such as carbon dioxide or chlorides is mainly responsible for the deterioration of reinforced concrete structures [4–6]. For new buildings exposed to natural environments, the durability of concrete elements can be guaranteed with proper mix design and construction details; when the environmental exposure is severe (i.e., industrial environments) or in the case of existing structures with durability deficiency, these problems could be solved by using electrochemical techniques, cathodic protection, corrosion inhibitors and surface treatments.

This Special Issue "Corrosion in Concrete: Inhibitors and Coatings" was launched to create new food for thought on the issue of durability of existing concrete structures and to collect the latest findings on the protection of reinforced concrete from corrosion of reinforcement and deterioration of cement matrix. Papers collected in this Special Issue are perfectly consistent with the proposed topic and, embracing different facets, provide the reader with different points of view. A brief summary is given below.

The paper by Coffetti et al. [7] reports—in an exhaustive and discriminating way—the main coating types available on the market, outlining a handbook to help in choosing the most suitable protective treatment as a function of the environment, the concrete properties and the job-site characteristics. The performance of polymeric and cementitious coatings, hydrophobic impregnations and pore blocking treatments were also analyzed in compliance with current standards and the different optimal uses for each protective were outlined from the comparison of the experimental results.

Jiang et al. [8] and Merachtsaki et al. [9] proposed innovative coatings for the protection of concrete against seawater and bio-corrosion, respectively. The first paper focuses on 85 geopolymer pastes containing fly ash, ground granulated blast-furnace slag, metakaolin and Portland cement able to ensure excellent adhesion to concrete (up to 3.4 MPa) and ultra-high resistance in seawater. The latter deals with the development of a magnesium hydroxide coating able to prevent the microbiologically induced corrosion of concrete sewer pipes. The prolonged protection of the coating was confirmed by using short and long duration accelerated sulfuric acid spraying tests and measuring the coating consumption with scanning electron microscope analyses, X-ray diffractions and attenuated total reflectance measurements.

Coppola et al. [10] analyzed the behavior of a silane-based water-repellent migrant corrosion inhibitor applied on different concretes, focusing on the influence of the concrete composition (water-to-cement ratio, cement type, cement factor) and curing age on the effectiveness of the corrosion inhibitor. Experimental results evidenced the effectiveness of silane-based treatment in improving the durability of concrete exposed to chlorides both in accelerated and natural diffusion tests by means of a water repellent effect.

Several authors have delt with the corrosion of the reinforcement bars in presence of chlorides, treating the subject from different points of view. Jasniok et al. [11,12] investigated the corrosion behavior and the bond strength of zinc-coated low-carbon reinforcing steel in chloride contaminated concrete in comparison with black steel. Experimental results evidenced the favorable impact of zinc coating on steel by providing two-year protection against corrosion in environments with high chloride content. On the contrary, it was noticed that the presence of zinc coating had a deleterious impact on the parameters of anchorage. Zhao et al. [13] proposed to pretreat the steel fiber with zinc phosphate to improve the resistance of fiber reinforced concrete against chlorides. It was evidenced that the phosphating treatment of fibers allows to enhance the bond strength between steel and concrete and improves the corrosion resistance of fibers without affecting their surface morphology. Finally, the paper by Garcia-Contreras et al. [14] studied the effect of impressed current cathodic protection on reinforced concrete manufactured with fly ash as cement replacement. Results evidenced that the microstructure of cementitious matrix strongly influenced the corrosion behavior of steel in concrete.

Funding: This research received no external funding.

Conflicts of Interest: The authors declare no conflict of interest.

References

1. Chen, C.; Habert, G.; Bouzidi, Y.; Jullien, A. Environmental impact of cement production: Detail of the different processes and cement plant variability evaluation. *J. Clean. Prod.* **2010**, *18*, 478–485. [CrossRef]
2. Damtoft, J.S.; Lukasik, J.; Herfort, D.; Sorrentino, D.; Gartner, E.M. Sustainable development and climate change initiatives. *Cem. Concr. Res.* **2008**, *38*, 115–127. [CrossRef]
3. Coppola, L.; Coffetti, D.; Crotti, E.; Gazzaniga, G.; Pastore, T. An Empathetic Added Sustainability Index (EASI) for cementitious based construction materials. *J. Clean. Prod.* **2019**, *220*, 475–482. [CrossRef]
4. Bertolini, L.; Elsener, B.; Pedeferri, P.; Redaelli, E.; Polder, R.B. *Corrosion of Steel in Concrete: Prevention, Diagnosis, Repair*; Wiley WCH: Hoboken, NJ, USA, 2013; ISBN 978-3-527-33146-8.
5. Zacchei, E.; Nogueira, C.G. Chloride diffusion assessment in RC structures considering the stress-strain state effects and crack width influences. *Constr. Build. Mater.* **2019**, *201*, 100–109. [CrossRef]
6. Wu, L.; Wang, Y.; Wang, Y.; Ju, X.; Li, Q. Modelling of two-dimensional chloride diffusion concentrations considering the heterogeneity of concrete materials. *Constr. Build. Mater.* **2020**, *243*, 118213. [CrossRef]
7. Coffetti, D.; Crotti, E.; Gazzaniga, G.; Gottardo, R.; Pastore, T.; Coppola, L. Protection of Concrete Structures: Performance Analysis of Different Commercial Products and Systems. *Materials* **2021**, *14*, 3719. [CrossRef] [PubMed]
8. Jiang, C.; Wang, A.; Bao, X.; Chen, Z.; Ni, T.; Wang, Z. Protective Geopolymer Coatings Containing Multi-Componential Precursors: Preparation and Basic Properties Characterization. *Materials* **2020**, *13*, 3448. [CrossRef] [PubMed]
9. Merachtsaki, D.; Tsardaka, E.-C.; Anastasiou, E.; Zouboulis, A. Evaluation of the Protection Ability of a Magnesium Hydroxide Coating against the Bio-Corrosion of Concrete Sewer Pipes, by Using Short and Long Duration Accelerated Acid Spraying Tests. *Materials* **2021**, *14*, 4897. [CrossRef] [PubMed]
10. Coppola, L.; Coffetti, D.; Crotti, E.; Gazzaniga, G.; Pastore, T. Chloride Diffusion in Concrete Protected with a Silane—Based Corrosion Inhibitor. *Materials* **2020**, *13*, 2001. [CrossRef] [PubMed]
11. Jaśniok, M.; Kołodziej, J.; Gromysz, K. An 18-Month Analysis of Bond Strength of Hot-Dip Galvanized Reinforcing Steel B500SP and S235JR+AR to Chloride Contaminated Concrete. *Materials* **2021**, *14*, 747. [CrossRef] [PubMed]
12. Jaśniok, M.; Sozańska, M.; Kołodziej, J.; Chmiela, B. A Two-Year Evaluation of Corrosion-Induced Damage to Hot Galvanized Reinforcing Steel B500SP in Chloride Contaminated Concrete. *Materials* **2020**, *13*, 3315. [CrossRef] [PubMed]
13. Zhao, X.; Liu, R.; Qi, W.; Yang, Y. Corrosion resistance of concrete reinforced by zinc phosphate pretreated steel fiber in the presence of chloride ions. *Materials* **2020**, *13*, 3636. [CrossRef] [PubMed]
14. García-Contreras, J.; Gaona-Tiburcio, C.; López-Cazares, I.; Sanchéz-Díaz, G.; Ibarra Castillo, J.C.; Jáquez-Muñoz, J.; Nieves-Mendoza, D.; Maldonado-Bandala, E.; Olguín-Coca, J.; López-León, L.D.; et al. Effect of Cathodic Protection on Reinforced Concrete with Fly Ash Using Electrochemical Noise. *Materials* **2021**, *14*, 2438. [CrossRef] [PubMed]

Article

Chloride Diffusion in Concrete Protected with a Silane-Based Corrosion Inhibitor

Luigi Coppola [1,2,3,*], **Denny Coffetti** [1,2,3], **Elena Crotti** [1,2,3], **Gabriele Gazzaniga** [1,3] and **Tommaso Pastore** [1,2,3]

[1] Department of Engineering and Applied Sciences, University of Bergamo, 24044 Dalmine (BG), Italy; denny.coffetti@unibg.it (D.C.); elena.crotti@unibg.it (E.C.); gabriele.gazzaniga@unibg.it (G.G.); tommaso.pastore@unibg.it (T.P.)

[2] UdR Materials and Corrosion, Consorzio INSTM, 50121 Florence, Italy

[3] UdR Bergamo, Consorzio CSGI, 50019 Sesto Fiorentino, Italy

* Correspondence: luigi.coppola@unibg.it; Tel.: +39-035-2052054

Received: 25 March 2020; Accepted: 23 April 2020; Published: 24 April 2020

Abstract: One of the most important parameters concerning durability is undoubtedly represented by cement matrix resistance to chloride diffusion in environments where reinforced concrete structures are exposed to the corrosion risk induced by marine environment or de-icing salts. This paper deals with protection from chloride ingress by a silane-based surface-applied corrosion inhibitor. Results indicated that the corrosion inhibitor (CI) allows to reduce the penetration of chloride significantly compared to untreated specimens, independently of w/c, cement type, and dosage. Reduction of chloride diffusion coefficient (D_{nssn}) measured by an accelerated test in treated concrete was in the range 30–60%. Natural chloride diffusion test values indicate a sharp decrease in apparent diffusion coefficient (D_{app}) equal to about 75% when concrete is protected by CI. Mechanism of action of CI in slowing down the chloride penetration inside the cement matrix is basically due to the water repellent effect as confirmed by data of concrete bulk electrical resistivity.

Keywords: durability of concrete; chloride penetration; rebar corrosion; corrosion inhibitor; silane-based surface treatment

---

1. Introduction

Concrete alkalinity promotes the formation of a passive protective oxide layer able to prevent corrosion of steel rebars and guarantees an adequate service life of reinforced concrete structures [1]. However, de-passivation of reinforcements can take place for many reasons, among which the most widespread is when chlorides reach a critical concentration at the interface cement matrix/steel bar. Chlorides can penetrate inside the cement matrix from external sources by capillary suction or by diffusion, for example, from contact or proximity to sea water or in a structure where de-icing salts are used, but can also be added incorrectly into the concrete through contaminated aggregates, admixtures, or water [2,3]. It is well known that the chloride-induced corrosion is one of the most dangerous and common phenomena for reinforced concrete structures in the marine environment or exposed to de-icing salts [4]. In a perspective of sustainability in the construction sector and to prevent premature structural failures due to chloride-induced corrosion, it is important to investigate possible strategies to counteract this degradation phenomenon [5,6]. Before dealing with these preventive methods, it is important to underline how the correct choice of concrete cover and mixture composition plays an important role in hindering the diffusion process of chlorides inside the cement matrix [7,8]. In agreement with the diagram of Tuuti [9], one of the main goals consists in slowing down the chloride diffusion inside the matrix in order to delay the onset of the corrosion process. Several alternative strategies have been proposed for increasing the durability of reinforced concrete structures exposed to chloride-rich environments such as coatings [10,11], cathodic protection [12,13], chloride extraction [14], and use of corrosion inhibitors [15–19]. Among these, the use of corrosion inhibitors (CI) is one of the most effective and cheaper ways to prevent the chloride-induced corrosion of reinforced concrete structures.

Two different types of corrosion inhibitors are available on the market: the admixed inhibitors, added to fresh concrete, and migrating corrosion inhibitors—also called penetrating inhibitors or surface-applied corrosion inhibitors—applied on the hardened concrete surface [20]. In particular, the latter seems to be an interesting solution for existing concrete structures exposed to chlorides such as infrastructures, bridges, marine structures, seawater pipelines, and chemical industries [21]. Many investigations have been conducted on surface-applied corrosion inhibitors. Soylev et al. evidenced the effectiveness of amino alcohol-based surface-applied corrosion inhibitors due to a pore-blocking effect as demonstrated by the resistivity measurements of concrete [22]. However, the inhibitors seem to block the pores on the surface of concrete rather than the bulk concrete similarly to a waterproofing treatment [23]. Holloway et al. found that the corrosion inhibitor was still present in the concrete cover at 5 years from application [24]. Research by Fedrizzi et al. demonstrated that the simultaneous use of the alkanolamine-based inhibitor with a good barrier coating offers protection against chloride-induced rebar corrosion [25]. Finally, the efficiency of a surface-applied corrosion inhibitor based on alkylaminoalcohol was highlighted by Morris and Vazquez, especially when it was applied on low-quality concretes manufactured with raw materials contaminated with chloride ions [26].

The purpose of this paper is to evaluate the performances of a silane-based corrosion inhibitor applied on the surface of concrete element in order to slow down chloride diffusion in cement matrix and, consequently, to delay the onset of the corrosion process. The experimental program was carried out both in the form of accelerated and natural diffusion tests in different concrete mixtures manufactured in order to evaluate—other than the efficiency of CI treatment—the influence of w/c, cement type, and cement factor on the penetration mechanism. For each concrete, the chloride diffusion coefficient (D_{nssm} also called D_{RCM}) was calculated to quantify the reduction in the penetration of chloride into the cement matrix in accordance with Spiesz and Brouwers [27] and Li et al. [28].

2. Materials and Methods

Seven different types of concrete have been manufactured (Table 1). Water/cement ratio and cement factor were selected in order to meet requirements for the exposure classes XD and XS according to EN 206 [29]. The denomination of the different concretes was made taking into consideration the different variables analysed: type of cement (natural pozzolanic cement: CEM IV/A-P 42.5R, limestone Portland cement: CEM II/A-L 42.5R, and blast furnace cement: CEM III/A 42.5R—Table 2), w/c ratio (0.55, 0.50, and 0.45) and cement factor (320, 340, and 360 kg/m^3). Finally, natural siliceous sand and gravel (three different gradings) with a maximum size equal to 22 mm were combined in order to meet the Bolomey curve (Figure 1).

Table 1. Composition of the concretes.

Concrete	CEM IV/A-P 42.5 R [kg/m^3]	CEM II/A-LL 42.5 R [kg/m^3]	CEM III/A 42.5 R [kg/m^3]	Aggregates [kg/m^3]	Water [kg/m^3]	w/c
IV-0.55-320	320			1880	176	0.55
IV-0.50-340	340			1875	170	0.50
II-0.50-340		340		1885	170	0.50
III-0.50-340			340	1875	170	0.50
IV-0.50-320	320			1915	160	0.50
IV-0.50-360	360			1830	180	0.50
IV-0.45-360	360			1885	162	0.45

Table 2. Main properties of cements.

Properties	CEM IV/A-P 42.5 R	CEM II/A-LL 42.5 R	CEM III/A 42.5 R
Specific mass [kg/dm^3]	3.01	3.10	3.05
Specific surface [m^2/kg]	480	400	400
Setting time [min]	>130	>130	>60

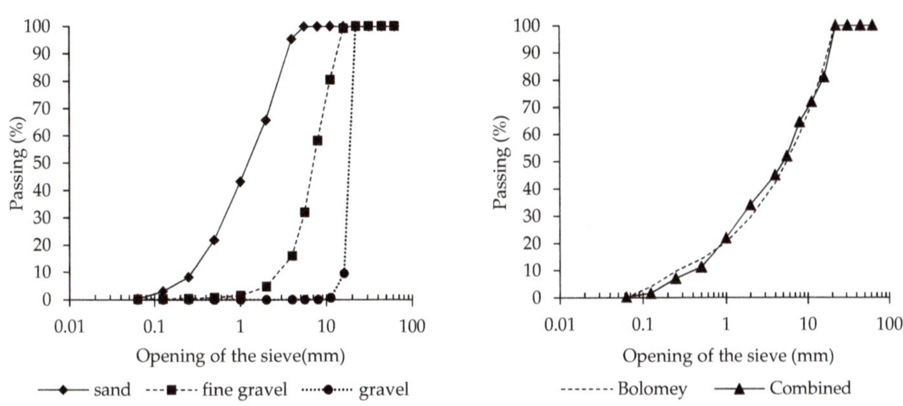

Figure 1. Grading curves of the aggregates (**left**). Bolomey and combined aggregate curves (**right**).

At the end of the mixing procedure, workability was measured by flow table according to EN 12350-5 [30]. In addition, specific mass and entrapped air were evaluated on fresh concretes according to EN 12350-6 [31] and EN 12350-7 [32] standards, respectively. For each concrete mixture, 30 cubic specimens (150 × 150 mm) and 32 cylindric specimens (d = 100 mm and h = 200 mm) were manufactured. Concrete samples were removed from the steel molds after 24 h and subsequently cured according to the scheme in Table 3. Compressive strength on hardened concrete was also determined at different ages (EN 12390-3 [33]).

Table 3. Specimens, curing procedure, and preparation.

Test	Curing and Preparation	Specimen Format	Note
Compressive strength	Curing at 20 °C and R.H. > 95% until the deadline	Cube 150 mm	1–7–28–70–100–130–210 days; 2 samples for each age
Accelerated chloride diffusion test - Bulk electrical resistivity test	Curing at 20 °C and R.H. > 95% for 7 days; Preparation of specimens by sawing and grinding; Drying in oven at 60 °C; Application of the CI; Water saturation for 24 h of samples for 7-day tests; Soaking the specimen in water for 28-day tests	Cylinder d: 100 mm h: 50 mm - d: 100 mm h: 100 mm	7–28 days; 8 samples for each age (4 treated and 4 untreated)
Natural chloride diffusion test	Curing at 20 °C and R.H. > 95% for 14 days; Curing at 20 °C and R.H. 60% for 28 days; Application of the CI; Immersion of specimens in a 3 wt.% NaCl solution until the deadline	Cube 150 mm	1–2–3–6 months of immersion; 4 samples for each age (2 treated and 2 untreated)

For the estimation of chloride penetration into concrete, accelerated migration tests, natural diffusion tests, and bulk electrical resistivity measurements were carried out on concrete with and without silane-based surface treatment as detailed in the following paragraphs. In particular, after a proper curing time (Table 3), half of the specimens were subjected to a silane-based surface-applied corrosion inhibitor while the others were used as an untreated reference. The properties of the corrosion inhibitor and application procedure are reported in Table 4.

Table 4. Properties of the corrosion inhibitor and application procedure.

Properties	Value
Color	Straw yellow
Viscosity [mPa·s]	0.95 ± 0.05
Dry residue [%]	7 ± 0.3
pH	6.5 ± 0.2
Density [kg/dm^3]	0.88 ± 0.05
Average consumption [L/m^2]	0.25 for each coat
Number of coats	4
Time between coats	15 min
Application method	Brush

2.1. Accelerated Chloride Migration Tests

Accelerated chloride migration tests were carried out according to NT BUILD 492 [34]. A cylindrical water-saturated concrete specimen (100 mm diameter and 50 mm height) was placed between two cells, one of them filled with 0.30 N NaOH solution and the other with a 10 wt.% NaCl solution. A 30 V DC potential was applied across the sample and the initial current was evaluated. Based on the measured initial current, the test voltage and the test duration were selected according to the NT BUILD 492. A data logger (Germann Instruments Ltd., Copenhaghen, Denmark) was used to record the electrical current, the temperature, and the electrical permeability during the test. Finally, the penetration depth of chlorides was determined by means of a 0.1 M silver nitrate solution [35,36] and the chloride diffusion coefficient (D_{nssm}) was calculated by the following equation [27]:

$$D_{nssm} = \frac{RT}{zFE} \times \frac{x_d - \alpha \sqrt{x_d}}{t} \tag{1}$$

with

$$E = \frac{U - 2}{L}, \tag{2}$$

$$\alpha = 2 \sqrt{\frac{RT}{zFE}} \, erf^{-1}\left(1 - \frac{2c_d}{c_0}\right) \tag{3}$$

where z is the absolute value of ion valence, F is the Faraday constant, U is the absolute value of the applied voltage, R is the gas constant, T is the average value of the initial and final temperatures in the anolyte solution, L is the thickness of the specimen, X_d is the average value of the penetration depths, t is the test duration, erf^{-1} is the inverse of error function, C_d is the chloride concentration at which the color changes, and C_0 is the chloride concentration in the catholyte solution.

2.2. Natural Chloride Diffusion Tests

For the natural chloride diffusion test, 150 mm cubic specimens were stored in a 3 wt.% NaCl solution at 20 °C for six months. The solution was replaced monthly and, at fixed intervals (1–2–3–6 months), the samples were split into two halves by means of a compression testing machine (Controls Spa, Liscate (MI), Italy) and the penetration of chlorides was measured using the previously described colorimetric method based on silver nitrate [35,36].

2.3. Bulk Electrical Resistivity Tests

The standard method reported in ASTM C1760 was used to evaluate the bulk electrical resistivity of concrete with and without corrosion inhibitor. The water-saturated concrete sample (100 mm diameter and 100 mm height) was positioned between the test cells used for accelerated chloride migration test containing 3 wt.% NaCl solution and an electrical potential of 60 V DC was applied across the specimen. The bulk electrical resistivity was calculated using the following equation:

$$\rho = \frac{V}{I} \times \frac{\pi d^2}{4L} \tag{4}$$

where ρ is the electrical resistivity in $k\Omega \cdot cm$, V is the applied voltage (60 V), I is the current in A, d is the specimen diameter (100 mm), and L is the specimen length (100 mm).

3. Results and Discussion

3.1. Fresh Properties

In Table 5, the fresh properties of concretes are listed. No substantial differences between the different mixtures in terms of workability, air content, and specific mass at fresh state were noticed. All concretes evidenced workability class F4 according to EN 206 [29] and the air content reflects the one expected for concrete manufactured with aggregate having maximum size equal to 22 mm. Finally, the specific mass at fresh state is similar for all concretes investigated.

Table 5. Properties of concretes at fresh state.

Concrete	Workability [mm]	Air Content [%]	Specific Mass [kg/m^3]
IV-0.55-320	550	1.6	2375
IV-0.50-340	520	1.7	2380
II-0.50-340	530	1.8	2395
III-0.50-340	540	1.9	2385
IV-0.50-320	530	1.9	2395
IV-0.50-360	530	1.8	2375
IV-0.45-360	510	1.6	2405

3.2. Elasto-Mechanical Properties

Table 6 shows results of compressive strength at different ages; as expected, the lower the w/c, the higher the compressive strength values.

Table 6. Cubic compressive strength (fc) results.

Concrete	w/c Ratio	Cubic Compressive Strength: fc [MPa]							210 d-fc/28 d-fc
		1 d	7 d	28 d	70 d	100 d	130 d	210 d	
IV-0.55-320	0.55	11.1	23.1	32.0	36.6	39.4	39.9	41.2	129%
IV-0.50-340	0.50	13.5	28.4	35.8	40.8	42.5	43.5	46.4	130%
II-0.50-340	0.50	16.7	28.8	36.8	39.8	41.4	43.1	43.4	118%
III-0.50-340	0.50	13.5	25.3	32.3	38.1	40.9	41.9	42.9	133%
IV-0.50-320	0.50	13.8	29.0	35.8	39.7	43.0	43.3	45.0	126%
IV-0.50-360	0.50	17.5	31.3	36.8	40.5	43.7	45.0	46.9	127%
IV-0.45-360	0.45	17.8	32.8	41.8	45.8	48.8	50.0	53.1	127%

Compressive strength at 210 days is 26–30% higher than the corresponding value achieved at 28 days for concretes manufactured with pozzolanic (IV) and blastfurnace (III) cements. The 210-day strength value of limestone Portland cement concrete (II), on the contrary, is only 18% higher than the 28-day compressive strength. Data confirm that when a pozzolanic or blastfurnace cement is used, a higher increase of compressive strength with time is achieved as a consequence of the pozzolanic reaction [37].

3.3. Bulk Electrical Resistivity Tests Results

The average bulk electrical resistivity of water-saturated concretes is reported in Table 7. As shown, all values related to untreated concretes are in the range of 5–8 kΩ·cm after 7 days and 8–14 kΩ·cm after 28 days, in accordance with Layssi et al. [38] and Neville [37]. Small differences are detected by varying the cement factor at equal w/c; on the contrary, the electrical resistivity increase when low w/c was adopted and it decreases when limestone Portland cement (II) was used instead of pozzolanic cement (IV) or blastfurnace (III) cement due to the denser structure promoted by the pozzolanic reaction of slag and natural pozzolan [39]. The use of a surface-applied corrosion inhibitor on concrete determines a strong increase in electrical resistivity, both at 7 and 28 days. However, the increasing in electrical resistivity is higher at 7 days (about +85%–+145%) respect to that at 28 days (about +40%–+65%).

Table 7. Bulk electrical resistivity tests results.

Concrete	Bulk Electrical Resistivity at 7 d [kΩ·cm]		Bulk Electrical Resistivity at 28 d [kΩ·cm]	
	Untreated	Treated	Untreated	Treated
IV-0.55-320	5.6	11.3	9.2	15.0
IV-0.50-340	6.9	15.4	10.5	16.6
II-0.50-340	5.0	9.5	7.7	12.0
III-0.50-340	7.6	13.3	13.8	19.8
IV-0.50-320	6.2	15.2	12.8	18.6
IV-0.50-360	7.2	17.2	12.9	18.3
IV-0.45-360	7.9	17.1	12.0	18.5

Figure 2 shows the bulk electrical resistivity of both treated and untreated specimens as a function of compressive strength of concrete. Figure 3 clearly confirms a significant increase in the electrical resistivity as a consequence of the surface treatment by CI and it seems to indicate that the surface-applied corrosion inhibitors acts as a water repellent protection.

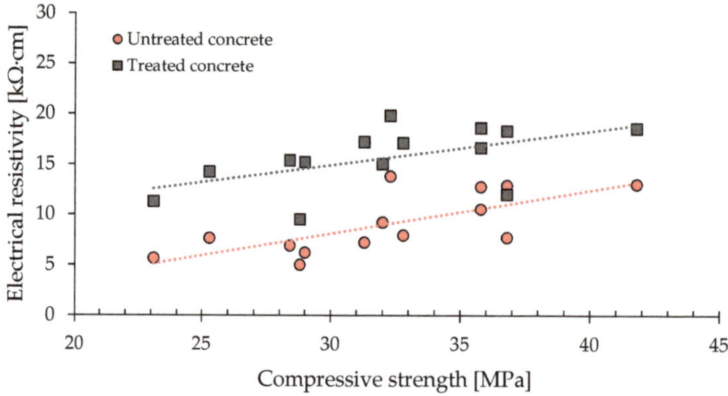

Figure 2. Correlation between electrical resistivity and compressive strength for treated and untreated concrete.

Data highlight the positive role of w/c since the electrical resistivity increases with concrete compressive strength independently of whether the specimen is treated or not, confirming results available in literature [40–42]. Moreover, the slope of the trend line for treated specimens is higher than that of untreated concrete. Assuming the strong direct relationship between the electrical resistivity and chloride diffusion reported in several papers [43–45], data reported in Figure 3 indicate that the corrosion inhibitor is more effective in slowing down chloride diffusion in concretes having high mechanical performances.

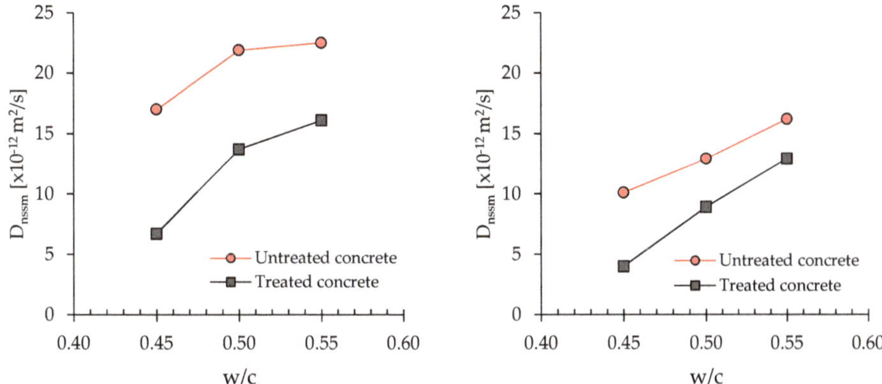

Figure 3. Chloride diffusion coefficient (D_{nssn}) vs. w/c ratio at 7 days (**left**) and 28 days (**right**).

3.4. Accelerated Chloride Migration Tests Results

The chloride diffusion coefficient (D_{nssn}) resulting from the accelerated diffusion test basically depends on the depth of chloride penetration in concrete (Table 8). As expected, the concretes manufactured with pozzolanic cement (IV) or blastfurnace cement (III) show a lower chloride diffusion coefficient with respect to limestone Portland cement-based mixtures (II). In particular, the D_{nssm} is in the range of $14-22 \times 10^{-12}$ m^2/s at 7 days and $6-16 \times 10^{-12}$ m^2/s at 28 days for III and IV samples, while II specimens reach values close to 28×10^{-12} m^2/s and 21×10^{-12} m^2/s, respectively. Protecting the concrete surface by the CI treatment determines a significant reduction of chloride penetration, independently of the age of concrete (7 or 28 days) when the accelerated diffusion test is carried out. The reduction of D_{nssm} is close to 30–40% if measured on samples water cured for 7 days and it slightly decreases at 21–39% when concrete is cured 28 days.

Table 8. Values of chloride diffusion coefficient of concretes.

Concrete	Untreated Specimens		Treated Specimens		Reduction [%]	
	7 d	28 d	7 d	28 d	7 d	28 d
IV-0.55-320	22.5	16.1	16.4	12.9	27.1	21.3
IV-0.50-340	21.9	13.7	12.9	8.9	41.1	35.1
II-0.50-340	28.1	20.8	19.5	14.0	30.8	32.7
III-0.50-340	14.0	7.6	9.6	5.7	31.4	25.0
IV-0.50-320	21.6	13.1	13.2	9.8	38.9	25.2
IV-0.50-360	18.7	12.6	12.7	9.2	32.1	27.0
IV-0.45-360	17.0	6.7	10.1	4.0	40.6	39.4

The header spans D_{nssm} [$\times 10^{-12}$ m^2/s].

Figure 3 reports the chloride diffusion coefficient (D_{nssn}) of pozzolanic cement-based concretes (IV) as a function of w/c ratio. Data are in good agreement with electrical resistivity results confirming the effectiveness of CI treatment in preventing chloride ingress inside the matrix. Moreover, according to electrical resistivity data, the efficiency of the CI treatment seems to be higher than the lower the w/c.

Figure 4 presents the chloride diffusion coefficient (D_{nssn}) vs. cement factor for concrete manufactured with CEM IV/A-P 42.5 R at the same w/c (0.50). Results confirm the positive role of the CI treatment, independently of the cement dosage. Similar to the bulk electrical resistivity, the chloride diffusion coefficient is not strongly influenced by the cement factor. Experimental results are in agreement with Bertolini et al. [46], affirming the binder content of the cement-based mixtures does not entail significant differences in terms of resistance to the penetration of chlorides.

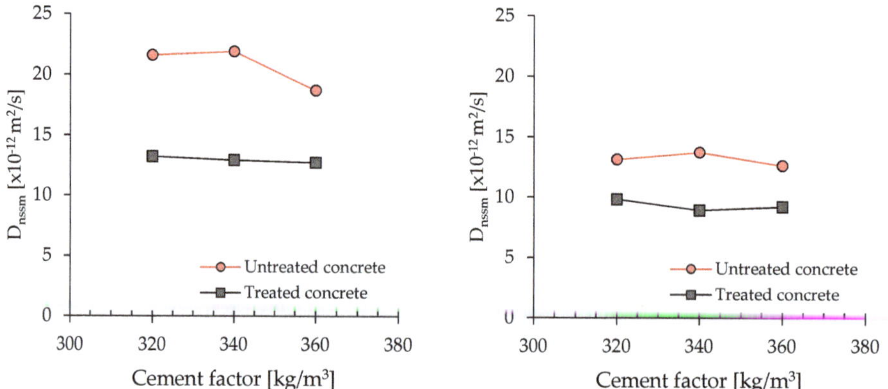

Figure 4. Chloride diffusion coefficient (D_{nssn}) vs. cement dosage at 7 days (**left**) and 28 days (**right**).

Figure 5 shows chloride diffusion coefficient (D_{nssn}) vs. cement type for concrete manufactured with the same w/c (0.50). Data confirm the efficiency of CI treatment independently of the cement type. Results also indicate the positive role of pozzolanic and blastfurnace cement in reducing the chloride penetration inside the matrix as a consequence of the binding capacity of pozzolanic reaction products [47–49].

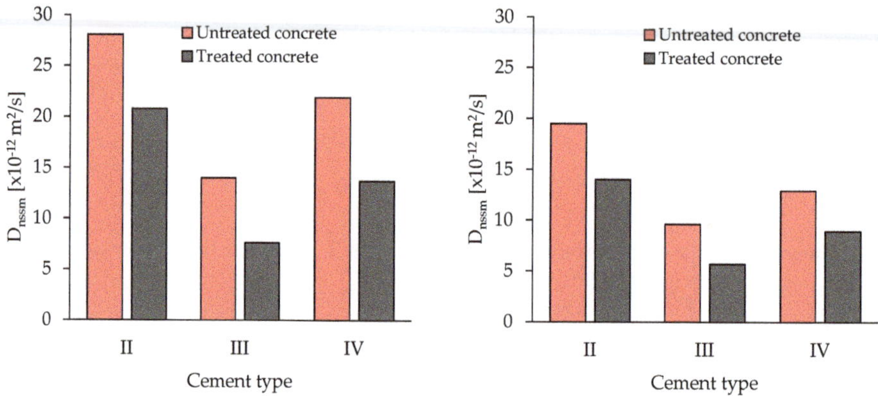

Figure 5. Chloride diffusion coefficient (D_{nssn}) vs. cement type for concrete manufactured with the same w/c (0.50) at 7 days (**left**) and 28 days (**right**).

On the basis of the experimental results of the accelerated chloride diffusion test, it is possible to affirm that the surface-applied corrosion inhibitor performs better if applied on concrete intrinsically resistant to chloride penetration manufactured with a pozzolanic or blastfurnace cement and with a low w/c ratio.

Finally, from the analysis of parameters resulting from the accelerated chloride migration tests and the bulk electrical resistivity tests, it was possible, in accordance with Layssi et al. [38], to correlate the chloride diffusion coefficient and the electrical conductivity of concrete (Figure 6). A linear correlation can be found, in accordance with the Nernst–Einstein equation, that can lead to hypothesize that the protective corrosion inhibitor acts only in terms of increasing electrical resistivity (water repellent effect) without chemically modifying the ability to bind chloride ions.

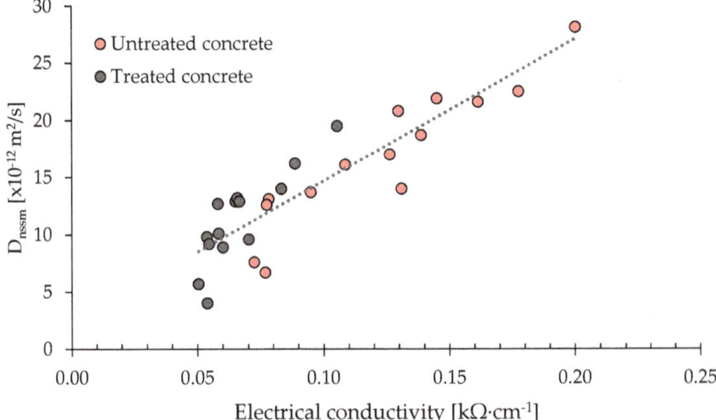

Figure 6. Correlation between chloride diffusion coefficient and electrical conductivity of concrete.

3.5. Natural Migration Test

Figure 7 shows chloride penetration vs. time for concrete specimens immersed in 3 wt.% NaCl aqueous solution. After 6 month of immersion, untreated samples evidenced a chloride penetration in the range of 11–20 mm, while the penetration depth of treated specimens is about 3–8 mm. Results clearly indicate that the CI treatment is strongly efficient in reducing the chloride diffusion independently of the w/c, the type and the dosage of cement, confirming the results registered for the accelerated chloride diffusion test.

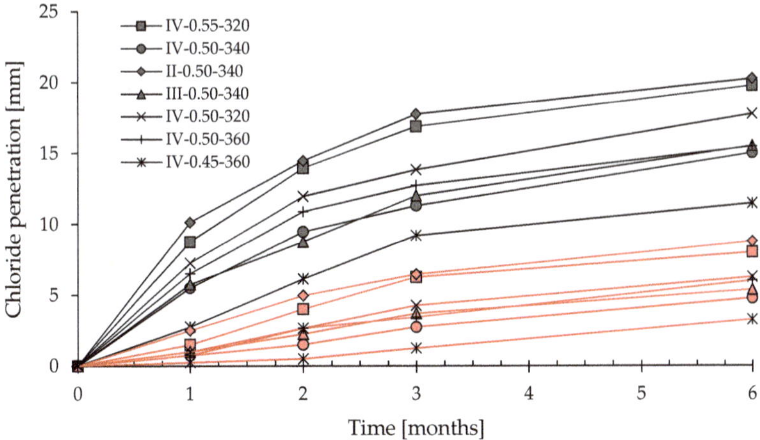

Figure 7. Correlation between chloride penetration and time (untreated concrete in black, treated concretes in red).

After three months of exposure, a reduction of 65–90% could be noticed in chloride penetration as a consequence of the CI treatment (Figure 8). After 6 months, concrete depth penetrated by chloride in treated samples is lower than that measured in untreated specimens of about 55–75%. Data seem to indicate that the efficiency of CI treatment in slowing down chloride diffusion slightly decreases with time. This behavior could be attributable to a partial leaching of the corrosion inhibitor as a consequence of the permanent immersion in chloride-based solution as already hypothesized by

Zheng et al. [50]. Further results at ages longer than 6 months are in progress to understand if efficiency of the surface-applied corrosion inhibitor remains constant or decreases with time.

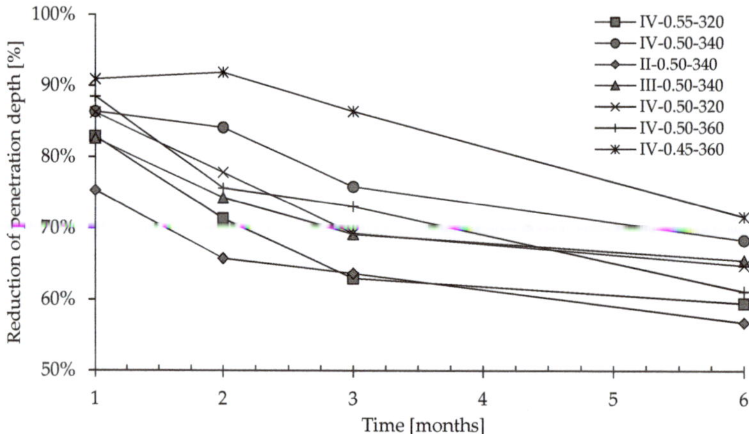

Figure 8. Reduction of depth penetration after natural migration test for different concretes.

Chloride penetration values (X) measured from the natural diffusion test over six months are used for the calculation of the average value (Table 9) of apparent diffusion coefficient (D_{app}) according to the following equation [51]:

$$0.66X = 1206 \sqrt{9.46 \times 10^7 \times T_{SLS} \times D_{app}} + dx \tag{5}$$

where 0.66 X coincides with the depth at which the critical concentration of chlorides is reached (0.4% respect to cement mass), T_{SLS} is the duration of the exposure to the chloride-rich solution expressed in years, and dx is the thickness of the convection layer depending on the concrete compressive strength. At equal w/c (0.50), D_{app} values are in the range of 0.20–0.30×10^{-12} m^2/s for untreated concretes manufactured with blastfurnace (III) or pozzolanic (IV) cements while limestone Portland cement-based mixtures (II) evidenced higher apparent diffusion coefficients, close to 1.15×10^{-12} m^2/s. The reduction in w/c promotes the formation of denser cementitious matrix with low D_{app} in accordance with the study of Neville [37].

Table 9. Average values of D_{app} for different concretes.

Concrete	D_{app} [$\times 10^{-12}$ m^2/s]		Reduction [%]
	Untreated Specimens	Treated Specimens	
IV-0.55-320	0.94	0.21	77.6
IV-0.50-340	0.19	0.04	78.9
II-0.50-340	1.15	0.26	77.4
III-0.50-340	0.32	0.07	78.1
IV-0.50-320	0.27	0.06	77.8
IV-0.50-360	0.24	0.06	75.0
IV-0.45-360	0.13	0.03	76.9

Treatment by the corrosion inhibitor determines a sharp decrease of D_{app}; values of treated concrete are in the range of 0.03–0.21×10^{-12} m^2/s, about 75% lower than those detected for concretes without treatment. The efficiency of CI treatment seems to be independent of w/c, type, and dosage of cement. However, the lowest values for D_{app} were obtained for those concretes intrinsically resistant to chloride penetration (low w/c and pozzolanic or blastfurnace cement), confirming the same results obtained for the accelerated chloride diffusion test.

Figure 9 compares D_{app} and 28-day D_{nssn}; the correspondence between the two coefficients is linear and the proportionality factor is consistent, as reported in study of Spiesz and Brouwers [27]. In particular, the regression line of treated specimens is placed below that of untreated concretes and the slope is lower compared to the same value of untreated samples. These two aspects confirm that the CI protective treatment is particularly effective since the increase of the D_{nssm} determines a slower growth of D_{app} in treated specimens.

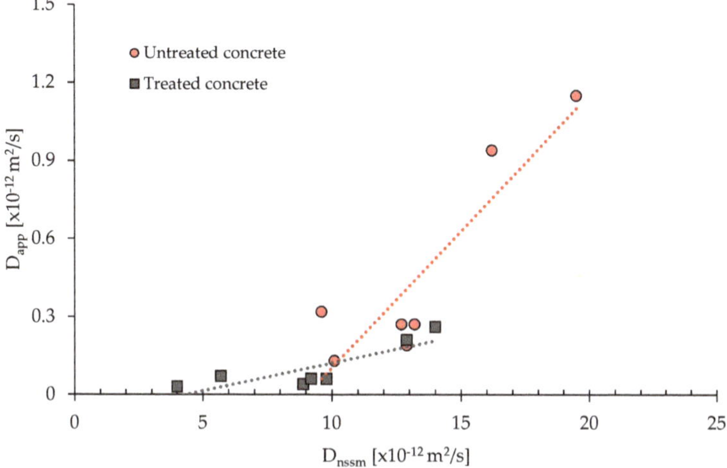

Figure 9. Correlation between D_{nssm} and D_{app}.

4. Conclusions

The following conclusions could be drawn from the present study:

- The surface-applied corrosion inhibitor allows to reduce significantly the penetration of chloride in concrete, independently of w/c, cement type, and dosage. Reduction of the chloride diffusion coefficient (D_{nssn}) measured by an accelerated test in treated concrete was in the range 30–60%. Natural chloride diffusion test values indicate a sharp decrease in D_{app} equal to about 75% when concrete is protected by the surface-applied CI.
- Mechanism of action of CI in slowing down the chloride penetration inside the cement matrix is basically due to the water repellent effect as confirmed by data of concrete electrical resistivity and accelerated chloride migration test results.
- The w/c ratio has confirmed to have a significant influence on chloride diffusion: the lower the w/c, the lower the penetration inside the cement matrix. On the contrary, no significant differences are observed in chloride penetration changing the cement dosage at the same w/c ratio.
- The type of cement considerably affects the chloride diffusion in concrete; in particular, it has been confirmed that limestone Portland cement (II) should be avoided in environments rich in chlorides, preferring pozzolanic (IV) or blast furnace (III) cements.

Further experimental data are in progress to evaluate the effectiveness of the surface-applied corrosion inhibitor at very long ages. Moreover, further studies should be focused on understanding if the migrating corrosion inhibitor is capable to stop corrosion in chloride contaminated concrete where the critical chloride concentration is reached on the steel rebars.

Author Contributions: Conceptualization, L.C., D.C., E.C. and G.G.; methodology, L.C., D.C., E.C. and G.G.; validation, D.C. and E.C.; investigation, G.G.; data curation, D.C.; writing—original draft preparation, G.G.; writing—review and editing, L.C. and D.C.; supervision, L.C. and T.P.; project administration, L.C. All authors have read and agreed to the published version of the manuscript.

Funding: This research received no external founding.

Acknowledgments: The authors would like to acknowledge Eng. Oscar Dozio for the development of the elasto-mechanical tests.

Conflicts of Interest: The authors declare no conflict of interest.

References

1. Bertolini, L.; Elsener, B.; Pedeferri, P.; Redaelli, E.; Polder, R.B. *Corrosion of Steel in Concrete: Prevention, Diagnosis, Repair*; Wiley WCH: Weinheim, Germany, 2013.

2. Collepardi, M.; Marcialis, A.; Turriziani, R. Penetration of Chloride Ions into Cement Pastes and Concretes. *J. Am. Ceram. Soc.* **1972**, *55*, 534–535. [CrossRef]

3. De Weerdt, K.; Colombo, A.; Coppola, L.; Justnes, H.; Geiker, M.R. Impact of the associated cation on chloride binding of Portland cement paste. *Cem. Concr. Res.* **2015**, *68*, 196–202. [CrossRef]

4. Wu, L.; Li, W.; Yu, X. Time-dependent chloride penetration in concrete in marine environments. *Constr. Build. Mater.* **2017**, *152*, 406–413. [CrossRef]

5. Coppola, L.; Coffetti, D.; Crotti, E.; Gazzaniga, G.; Pastore, T. An Empathetic Added Sustainability Index (EASI) for cementitious based construction materials. *J. Clean. Prod.* **2019**, *220*, 475–482. [CrossRef]

6. Coppola, L.; Coffetti, D.; Crotti, E. An holistic approach to a sustainable future in concrete construction. *IOP Conf. Ser. Mater. Sci. Eng.* **2018**, *442*, 012024. [CrossRef]

7. Lollini, F.; Carsana, M.; Gastaldi, M.; Redaelli, E.; Bertolini, L. The challenge of the performance-based approach for the design of reinforced concrete structures in chloride bearing environment. *Constr. Build. Mater.* **2015**, *79*, 245–254. [CrossRef]

8. Zhang, H.; Zhang, W.; Gu, X.; Jin, X.; Jin, N. Chloride penetration in concrete under marine atmospheric environment – analysis of the influencing factors. *Struct. Infrastruct. Eng.* **2016**, *12*, 1428–1438. [CrossRef]

9. Tuutti, K. *Corrosion of Steel in Concrete*; Swedish Foundation for Concrete Research: Stockholm, Sweden, 1982.

10. Brenna, A.; Beretta, S.; Berra, M.; Diamanti, M.V.; Ormellese, M.; Pastore, T.; Pedeferri, M.P.; Bolzoni, F. Effect of polymer modified cementitious coatings on chloride-induced corrosion of steel in concrete. *Struct. Concr.* **2019**, 201900255. [CrossRef]

11. Diamanti, M.V.; Brenna, A.; Bolzoni, F.; Berra, M.; Pastore, T.; Ormellese, M. Effect of polymer modified cementitious coatings on water and chloride permeability in concrete. *Constr. Build. Mater.* **2013**, *49*, 720–728. [CrossRef]

12. Oleiwi, H.M.; Wang, Y.; Curioni, M.; Chen, X.; Yao, G.; Augusthus-Nelson, L.; Ragazzon-Smith, A.H.; Shabalin, I. An experimental study of cathodic protection for chloride contaminated reinforced concrete. *Mater. Struct. Constr.* **2018**, *51*, 1–11. [CrossRef]

13. Wang, F.; Xu, J.; Xu, Y.; Jiang, L.; Ma, G. A comparative investigation on cathodic protections of three sacrificial anodes on chloride-contaminated reinforced concrete. *Constr. Build. Mater.* **2020**, *246*, 1–10. [CrossRef]

14. Tissier, Y.; Bouteiller, V.; Marie-Victoire, E.; Joiret, S.; Chaussadent, T.; Tong, Y.Y. Electrochemical chloride extraction to repair combined carbonated and chloride contaminated reinforced concrete. *Electrochim. Acta* **2019**, *317*, 486–493. [CrossRef]

15. Ormellese, M.; Berra, M.; Bolzoni, F.; Pastore, T. Corrosion inhibitors for chlorides induced corrosion in reinforced concrete structures. *Cem. Concr. Res.* **2006**, *36*, 536–547. [CrossRef]

16. Cabrini, M.; Fontana, F.; Lorenzi, S.; Pastore, T.; Pellegrini, S. Effect of Organic Inhibitors on Chloride Corrosion of Steel Rebars in Alkaline Pore Solution. *J. Chem.* **2015**, *2015*, 521507. [CrossRef]

17. Gartner, N.; Kosec, T.; Legat, A. The efficiency of a corrosion inhibitor on steel in a simulated concrete environment. *Mater. Chem. Phys.* **2016**, *184*, 31–40. [CrossRef]

18. Pan, C.; Li, X.; Mao, J. The effect of a corrosion inhibitor on the rehabilitation of reinforced concrete containing sea sand and seawater. *Materials* **2020**, *13*, 1480. [CrossRef]

19. Ngala, V.T.; Page, C.L.; Page, M.M. Corrosion inhibitor systems for remedial treatment of reinforced concrete. Part 2: Sodium monofluorophosphate. *Corros. Sci.* **2003**, *45*, 1523–1537. [CrossRef]

20. Söylev, T.A.; Richardson, M.G. Corrosion inhibitors for steel in concrete: State-of-the-art report. *Constr. Build. Mater.* **2008**, *22*, 609–622. [CrossRef]
21. Elsener, B.; Angst, U. 14—Corrosion inhibitors for reinforced concrete. In *Science and Technology of Concrete Admixtures*; Aïtcin, P.-C., Flatt, R.J., Eds.; Woodhead Publishing: Cambridge, UK, 2016; pp. 321–339. [CrossRef]
22. Söylev, T.A.; McNally, C.; Richardson, M. Effectiveness of amino alcohol-based surface-applied corrosion inhibitors in chloride-contaminated concrete. *Cem. Concr. Res.* **2007**, *37*, 972–977. [CrossRef]
23. Söylev, T.A.; McNally, C.; Richardson, M.G. The effect of a new generation surface-applied organic inhibitor on concrete properties. *Cem. Concr. Compos.* **2007**, *29*, 357–364. [CrossRef]
24. Holloway, L.; Nairn, K.; Forsyth, M. Concentration monitoring and performance of a migratory corrosion inhibitor in steel-reinforced concrete. *Cem. Concr. Res.* **2004**, *34*, 1435–1440. [CrossRef]
25. Fedrizzi, L.; Azzolini, F.; Bonora, P.L. The use of migrating corrosion inhibitors to repair motorways' concrete structures contaminated by chlorides. *Cem. Concr. Res.* **2005**, *35*, 551–561. [CrossRef]
26. Morris, W.; Vázquez, M. A migrating corrosion inhibitor evaluated in concrete containing various contents of admixed chlorides. *Cem. Concr. Res.* **2002**, *32*, 259–267. [CrossRef]
27. Spiesz, P.; Brouwers, H.J.H. The apparent and effective chloride migration coefficients obtained in migration tests. *Cem. Concr. Res.* **2013**, *48*, 116–127. [CrossRef]
28. Li, L.Y.; Easterbrook, D.; Xia, J.; Jin, W.L. Numerical simulation of chloride penetration in concrete in rapid chloride migration tests. *Cem. Concr. Compos.* **2015**, *63*, 113–121. [CrossRef]
29. EN 206-1. *Concrete. Specification, Performance, Production and Conformity*; BSI Standards Publication: London, UK, 2016.
30. EN 12350-5. *Testing Fresh Concrete. Flow Table Test*; BSI Standards Publication: London, UK, 2019.
31. EN 12350-6. *Testing Fresh Concrete. Density*; BSI Standards Publication: London, UK, 2019.
32. EN 12350-7. *Testing Fresh Concrete. Air Content. Pressure Methods*; BSI Standards Publication: London, UK, 2019.
33. EN 12390-3. *Testing Hardened Concrete. Compressive Strength of Test Specimens*; BSI Standards Publication: London, UK, 2019.
34. NT BUILD 492. *Concrete, Mortar and Cement-based Repair Materials. Chloride Migration Coefficient from Non-stedy-state Migration Experiments*; NordTest: Taastrup, Denmark, 1999.
35. Yuan, Q.; Deng, D.; Shi, C.; de Schutter, G. Application of silver nitrate colorimetric method to non-steady-state diffusion test. *J. Cent. South Univ.* **2012**, *19*, 2983–2990. [CrossRef]
36. Baroghel-Bouny, V.; Belin, P.; Maultzsch, M.; Henry, D. AgNO3 spray tests: Advantages, weaknesses, and various applications to quantify chloride ingress into concrete. Part 2: Non-steady-state migration tests and chloride diffusion coefficients. *Mater. Struct.* **2007**, *40*, 783. [CrossRef]
37. Neville, A. *Properties of Concrete*; Pearson Education Limited: Harlow, UK, 2011.
38. Layssi, H.; Ghods, P.; Alizadeh, A.; Salehi, M. Electrical Resistivity of Concrete. *Concr. Int.* **2015**, 293–302.
39. Lübeck, A.; Gastaldini, A.L.G.; Barin, D.S.; Siqueira, H.C. Compressive strength and electrical properties of concrete with white Portland cement and blast-furnace slag. *Cem. Concr. Compos.* **2012**, *34*, 392–399. [CrossRef]
40. McCarter, W.J.; Starrs, G.; Chrisp, T.M. Electrical conductivity, diffusion, and permeability of Portland cement-based mortars. *Cem. Concr. Res.* **2000**, *30*, 1395–1400. [CrossRef]
41. Shi, C. Effect of mixing proportions of concrete on its electrical conductivity and the rapid chloride permeability test (ASTM C1202 or ASSHTO T277) results. *Cem. Concr. Res.* **2004**, *34*, 537–545. [CrossRef]
42. Abdelrahman, M.; Xi, Y. The effect of w/c ratio and aggregate volume fraction on chloride penetration in non-saturated concrete. *Constr. Build. Mater.* **2018**, *191*, 260–269. [CrossRef]
43. Polder, R.B. Critical chloride content for reinforced concrete and its relationship to concrete resistivity. *Mater. Corros.* **2009**, *60*, 623–630. [CrossRef]
44. Ramezanianpour, A.A.; Pilvar, A.; Mahdikhani, M.; Moodi, F. Practical evaluation of relationship between concrete resistivity, water penetration, rapid chloride penetration and compressive strength. *Constr. Build. Mater.* **2011**, *25*, 2472–2479. [CrossRef]
45. Gulikers, J. Theoretical considerations on the supposed linear relationship between concrete resistivity and corrosion rate of steel reinforcement. *Mater. Corros.* **2005**, *56*, 393–403. [CrossRef]

46. Bertolini, L.; Lollini, F.; Redaelli, E. Comparison of Resistance to Chloride Penetration of Different Types of Concrete through Migration and Ponding Tests. In *Modelling of Corroding Concrete Structures*; Andrade, C., Mancini, G., Eds.; Springer: Dordrecht, The Netherlands, 2011; pp. 125–135.

47. Argiz, C.; Moragues, A.; Menéndez, E. Use of ground coal bottom ash as cement constituent in concretes exposed to chloride environments. *J. Clean. Prod.* **2018**, *170*, 25–33. [CrossRef]

48. Yildirim, H.; Ilica, T.; Sengul, O. Effect of cement type on the resistance of concrete against chloride penetration. *Constr. Build. Mater.* **2011**, *25*, 1282–1288. [CrossRef]

49. Dellinghausen, L.M.; Gastaldini, A.L.G.; Vanzin, F.J.; Veiga, K.K. Total shrinkage, oxygen permeability, and chloride ion penetration in concrete made with white Portland cement and blast-furnace slag. *Constr. Build. Mater.* **2012**, *37*, 652–659. [CrossRef]

50. Zheng, H.; Li, W.; Ma, F.; Kong, Q. The effect of a surface-applied corrosion inhibitor on the durability of concrete. *Constr. Build. Mater.* **2012**, *37*, 36–40. [CrossRef]

51. Petcherdchoo, A. Time dependent models of apparent diffusion coefficient and surface chloride for chloride transport in fly ash concrete. *Constr. Build. Mater.* **2013**, *38*, 497–507. [CrossRef]

 *materials* 

Article

A Two-Year Evaluation of Corrosion-Induced Damage to Hot Galvanized Reinforcing Steel B500SP in Chloride Contaminated Concrete

Mariusz Jaśniok [1,*], **Maria Sozańska** [2], **Jacek Kołodziej** [1] and **Bartosz Chmiela** [2]

[1]   Faculty of Civil Engineering, Silesian University of Technology, 5 Akademicka, 44-100 Gliwice, Poland; jacek.kolodziej@polsl.pl

[2]   Faculty of Materials Engineering, Silesian University of Technology, 8 Krasińskiego, 40-019 Katowice, Poland; maria.sozanska@polsl.pl (M.S.); bartosz.chmiela@polsl.pl (B.C.)

*   Correspondence: mariusz.jasniok@polsl.pl

Received: 3 June 2020; Accepted: 21 July 2020; Published: 25 July 2020

Abstract: Corrosion-induced damage to concrete reinforced with bars is a serious problem regarding technical and economic aspects and strongly depends on used materials, corrosion environment, and service life. Tests described in this paper refer to a two-year evaluation of the effectiveness of protection provided by zinc-coated low-carbon reinforcing steel of grade B500SP in concrete against chloride corrosion. Performed tests were comparative and included measurements conducted on four groups of concrete test elements with dimensions of 40 mm × 40 mm × 140 mm reinforced with a bar having a diameter of ϕ8 mm. Particular groups were a combination of different types of concrete with or without chloride additives, with galvanized or black steel. Chlorides as $CaCl_2$ were added to the concrete mix in the amount of 3% of cement weight in concrete. Reinforced concrete specimens were periodically monitored within two years using the following techniques: linear polarization resistance (LPR) and electrochemical impedance spectroscopy (EIS). Polarization measurements were conducted in a three-electrode arrangement, in which a rebar in concrete served as a working electrode, stainless steel sheet was used as an auxiliary electrode, and $Cl^-/AgCl,Ag$ was a reference electrode. Comparative tests of changes in the density of corrosion current in concrete specimens without chloride additives basically demonstrated no development of corrosion, and possible passivation was expected in case of black steel. Higher densities of corrosion current were observed for galvanized steel during first days of testing. The reason was the dissolution of zinc after the contact with initially high pH of concrete pore solution. Six-month measurements demonstrated a higher density of corrosion current in concrete specimens with high concentration of chlorides, which unambiguously indicated corrosion in concrete reinforced with galvanized or black steel. Densities of corrosion current determined for selected specimens dramatically decreased after an 18-month interval in measurements. Corrosion was even inhibited on black steel as an insulating barrier of corrosion products was formed. The above observations were confirmed with structural studies using scanning electron microscopy (SEM) and energy dispersive spectroscopy (EDS) techniques. Results obtained from corrosion (LPR, EIS) and structural (SEM, EDS) tests on specimens of concrete reinforced with steel B500SP demonstrated a very favorable impact of zinc coating on steel by providing two-year protection against corrosion in the environment with very high chloride content.

Keywords: concrete; reinforcing steel; zinc coating; HDG; corrosion; chlorides; LPR; EIS; SEM; EDS

---

1. Introduction

Issues addressed to the durability of reinforced concrete structures have been much tested and described in scientific papers of an interdisciplinary nature [1,2]. Bonds formed between concrete and steel reinforcement are the main used specificity of reinforced concrete as the construction material [3,4]. Therefore, concrete in a building transmits main compressive stresses, and steel transmits tensile stresses [5]. Regarding durability in the corrosion environment, reinforced concrete should be identified with a steel electrode (half-cell) immersed in the highly alkaline pore solution filling the porous structure of concrete [6,7].

A combination of two different construction materials, being advantageous from the mechanical point of view, can create serious difficulties and complications related to the evaluation of the structure safety with reference to its durability [8–10]. The main problem is the lack of direct access to steel reinforcement covered with concrete or failure to make observations without measuring equipment. Carbon dioxide inducing concrete carbonation [7,11] and chlorides [12] present in, inter alia, deicing agents or saltwater are among the most serious and common factors creating corrosion risk to reinforcement. Mechanisms resulting in damage to a passive layer, which is naturally formed on steel surface in the highly alkaline pore solution of concrete, have been precisely studied using versatile methods of measurements [13]. Laboratory tests and tests performed on facilities have demonstrated that elements of structures do not show any external symptoms of developing corrosion of reinforcing steel often for a very long time [14]. The literature basically describes three stages of corrosion degradation of reinforced concrete elements. At the first stage, corrosive agents gradually penetrate the porous structure of concrete cover and diffuse towards rebars. At the second stage, when corrosive agents reach the reinforcement surface, the passive layer is decomposed and corrosion processes are initiated on the steel surface. It is just the third stage when the volume of corrosion products (rust) is gradually increasing, corrosion damage to reinforced concrete in the form of cracks is visible, and finally spalling of smaller or bigger pieces of the cover is observed [15]. It is caused by tensile stresses formed in the concrete cover as a result of an expansion of corrosion products of steel and simultaneously very low tensile strength of concrete [16].

Today effects of degradation of reinforced concrete structures can be reduced or prevented using various methods. They include cathodic protection of reinforcement which provides permanent and controlled over time reduction of steel reinforcement potential in concrete to the level, at which metal exhibits resistance to corrosion. However, the use of the above method has to be preceded with the advanced diagnostic of corrosion employing such electrochemical methods as measurement of corrosion potential [17], measurement of the density of corrosion current [18]—the most common methods of linear polarization resistance (LPR) [19], electrochemical impedance spectroscopy (EIS) [20] and galvanostatic pulse (GP) method [21]. Those diagnostic methods are necessary for switching on and maintaining the direct current for cathodic protection throughout the whole service life of the structure. Another solution for increasing the durability of reinforcement in concrete elements of structures is the application of rebars made of stainless steel [22]. This type of steel, apart from basic components, such as iron and carbon, always contains at least 12% of chromium. Reinforcement for concrete structures is usually manufactured from austenitic, ferritic, or duplex (austenitic ferritic) steel [23]. Reinforcing steel can be replaced with non-metallic reinforcement made of FRP (Fiber-Reinforced Polymer) composite rebars. This type of reinforcement eliminates the problem of electrochemical corrosion [24,25]. Currently, three variants of those rebars are used in the construction sector: CFRP—with carbon fibers, GFRP—with glass fibers [26], and AFRP—with aramid fibers. However, non-metallic reinforcement in the construction sector has so far been rarely used when compared to traditional steel reinforcement.

In fact, protective epoxy and zinc coating is the latest possibility of direct protection of reinforcing steel in concrete against corrosion. Epoxy coatings on steel reinforcement in concrete behave like a typical electric insulator, which as a plastic material acts as a hardly permeable barrier cutting off steel from corrosive agents present in the concrete cover [27]. Unfortunately, technological processes occurring when reinforcement is laid in the formwork, and the process of concreting can cause point defects in those coatings. Places, where defects were produced, often invisible during construction works, can form local centers of corrosion [28,29] with time and concrete exposure to an aggressive environment. Considering the theory of reinforced concrete and structural mechanics, locally weakened reinforcing steel will also affect the local drop in load-carrying capacity of a reinforced concrete element, and consequently will pose a risk to the structure safety.

Zinc coatings prepared in HDG (Hot Dip Galvanizing) process [30–33] are free from drawbacks of epoxy coatings. In that case, any damage caused during the transport of reinforcement or its placing in the formwork does not create any risk for anti-corrosion protection of rebars in concrete [34]. The reason for no adverse effects is the formation of Fe-Zn alloy coating during the HDG process, which creates a galvanic cell in the concrete pore solution. As zinc is less noble regarding electrochemical series of metals, reinforcing steel is still protected against corrosion even in case of its punching [35]. It is a so-called "second phase of protection", in which the zinc coating takes the role of an electrochemically dissolving electrode, and iron serves as the cathode. And impenetrable zinc coating insulates rebar from humidity (water) and oxygen during the first phase of protection. Studies indicate the thickness of zinc coating properly prepared in the HDG process, usually ranges between 50–300 µm [36]. However, the recommended thickness should not be smaller than 80 µm. The disadvantage of zinc coating is its possible corrosion after the contact with highly alkaline pore solution at pH < 11.4 and pH $\geq$ 13.3 [37,38]. This unfavorable occurrence is usually observed in the initial phase of bonding fresh concrete mix in the formwork, in which galvanized reinforcement is laid [39]. In this context it is crucial that there is only one-side limit of pH values for black reinforcing steel—pH < 11.8, below which corrosion develops. The effect of concrete carbonation on galvanized steel depends on the neutralisation degree of pore solution, in which CO_2 gas dissolves [40]. If carbon dioxide acidifying the pore solution reduces pH values below 11.4, then zinc dissolution is initiated. However, according to Reference [37], zinc corrosion is unimportant, just minimal in this situation. As in case of black steel, the most destructive corrosion is observed on the coating of chloride ions. For the zinc coating, the limit content of chlorides is greater [37,41] than the one determined for black steel—0.4% [42] with reference to cement weight in concrete. For example, the boundary level equal to 1.5% is found in Reference [43], whereas Reference [44] mentions the percentage range of 1%–2%. Other studies did not define the percentage values of limit concentration, but specified chloride content initiating corrosion of zinc coating on the reinforcement as 2.5× [45] or, according to Reference [46] 4×–5× greater than the limit concentration determined for black steel. Discussions about limit values of chloride concentration unfavorable for the zinc coating manifest the difficulty in specifying any representative value. Such a great spread of values depends on the various compositions of concrete, mortar and grout used in tests. Moreover, different types of additives in concrete, different thickness of zinc coating and variation in chemical and phase composition of the coating across its thickness also affect corrosion damage.

The majority of described studies on galvanized reinforcing steel embedded in concrete refers to short time periods, that is, a few or more than ten weeks. The authors of this paper decided to perform comparative tests on the effectiveness of protection against chloride corrosion of zinc-coated low-carbon reinforcing steel of grade B500SP in concrete within over ten months. Apart from fundamental electrochemical polarization tests employing LPR and EIS techniques, the concrete-coating-reinforcement contact was additionally evaluated after two years of measurements using scanning electron microscopy (SEM) and energy dispersive spectroscopy (EDS) techniques.

2. Materials

Twelve concrete test elements with dimensions of 40 mm × 40 mm × 140 mm, reinforced with single steel bars were prepared for tests. Two batches of standard concrete mixes were used to place the concrete [47]. In both cases the mix contained 489 kg/m^3 of CEM I 42.5—SR3/NA cement, 1669 kg/m^3 of sand and aggregate of grain size of 2–8 mm. Water-cement ratio (*w/c*) was 0.45. About 3% of $CaCl_2$ by cement weight, dissolved in batched water, was added to one of those mixes. In conclusion, six test elements were prepared from concrete without any additives, and another group of six specimens was prepared from concrete with chloride additives.

Each concrete test element was reinforced with a single ribbed rebar with a diameter of $\phi 8$ mm, made of steel of grade B500SP classified to steel class C of high ductility [48]. As claimed by the manufacturer, maximum values of additives to low-carbon steel for elements, in which content exceeds 0.5% should not be greater than 1.65% for Mn, 0.85% for Cu, and 0.6% for Si.

In accordance with the primary purpose of studies, half of rebars made of steel of grade B500SP, used as the reinforcement in concrete test elements, was coated in the HDG process. For that purpose, holders were welded to cleaned bars cut to 1.5 m long pieces to provide galvanizing in a vertical position. Rebars placed on plating racks were successively immersed in a degreasing, pickling and finally fluxing bath. Prior to the immersion in the galvanizing bath, rebars were dried in a dryer. Bathing in molten zinc, at a temperature of ca. 450 °C, lasted 120 s. The obtained zinc coatings on steel rebars had an average thickness of ca. 100 μm, which met standard requirements [49,50]. Before the tests, the thickness of the zinc coating was initially verified under the metallographic microscope Delta Optical MET-200-TRF (Delta Optical Ltd., Nowe Osiny, Poland).

Before concreting the specimens, galvanized or black steel rebars were cut into sections of 15 cm in length. Ends of each rebar were protected with shrink film against crevice corrosion in zones, where rebars expanded from concrete. Then, the rebars were stabilized in standard moulds for preparing specimens of masonry mortar, and then concreted. After two days from concreting, the specimens were stripped from the mould. Two weeks after concreting, polarization measurements of the reinforcement began.

In summary, six concrete test elements were reinforced with galvanized rebars, and other six with black steel rebars. As two concrete mixes were used during tests, the total number of twelve test elements was divided into four groups, each of them containing three identical specimens. To differentiate clearly those groups, different colors and symbols were assigned to each of them as shown schematically in Figure 1:

- blue color and *B* symbol—black steel in concrete without any additives,
- green color and *G* symbol—galvanized steel in concrete without any additives,
- red color and *B-Cl* symbol—black steel in concrete with chlorides,
- violet color and *G-Cl* symbol—galvanized steel in concrete with chlorides.

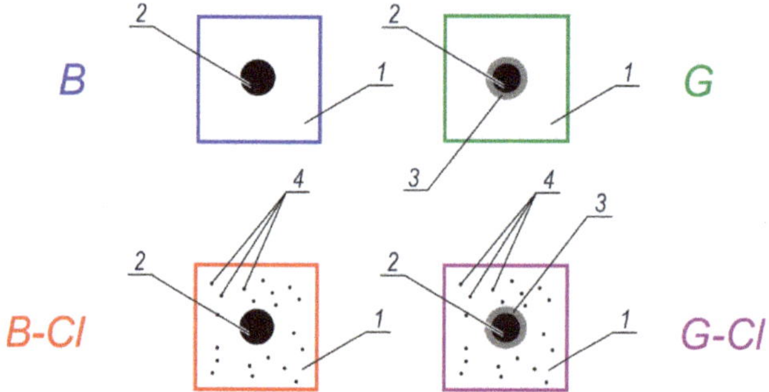

Figure 1. Four groups of reinforced concrete test elements used in tests: 1—concrete without additives, 2—black steel, 3—zinc coating on reinforcing steel, and 4—chlorides in concrete.

Assigned symbols and colors are consequently used in this paper to describe and analyze performed tests.

3. Test Methods

3.1. Electrochemical Polarization Tests

Polarization measurements of the reinforcement in concrete test elements were conducted in a three-electrode arrangement illustrated in Figure 2. Galvanized or black steel rebar 1 embedded in concrete 2 with or without chloride additives served as the working electrode. The auxiliary electrode 3 was made of a 2 mm thick stainless-steel sheet having the dimensions of 140 mm × 40 mm. A wet felt 4 with the dimensions of 140 cm × 40 cm provided the electric contact between the sheet 3 and the top surface of concrete. The reference electrode 5 (Cl^-/AgCl,Ag) was placed in a vertical ballast guide 6 to provide its stabilisation during tests. The ballast 6 was used to provide the proper adhesion of the auxiliary electrode 3 through the felt 4 to the top surface of the concrete specimen 2. All three electrodes were connected to the potentiostat 7 Gamry Reference 600.

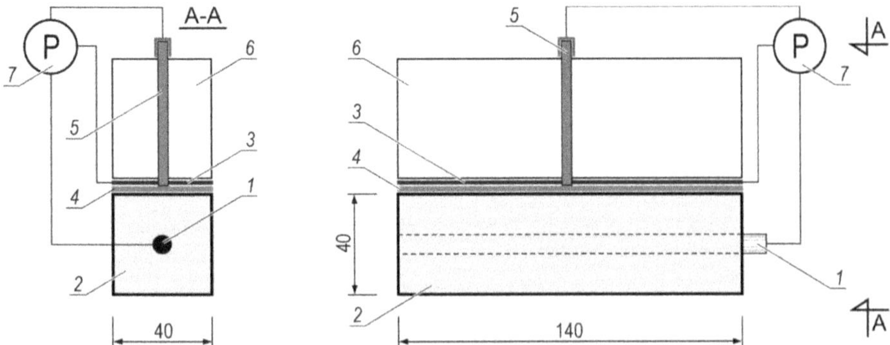

Figure 2. Measuring system for polarization tests of steel reinforcement in concrete: 1—working electrode, 2—concrete, 3—auxiliary electrode, 4—wet felt, 5—reference electrode Cl^-/AgCl,Ag, 6—ballast with a guide for reference electrode, 7—potentiostat.

Electrochemical tests were performed on the above measuring system using the following techniques: linear polarization resistance (LPR) and electrochemical impedance spectroscopy (EIS). The specimens were kept at ca. 20 ± 2 °C and the relative humidity of 50 ± 10%. Prior to beginning polarization tests on a given day, the top surface of the concrete specimen was immersed in tap water to the depth up to 5 ± 1 mm for ca. 30 min to ensure better conductivity of concrete cover in the tested reinforcement during measurements. After taking from water, the specimens were connected to the potentiostat 7 and changes of gradually stabilizing potential were observed with the reference electrode 5 for 60–120 min. When potential changes were at the level of 0.1 mV/s, EIS methods were performed on the steel reinforcement in concrete. Testing was performed at the fixed range of frequencies of 0.01 Hz–1 kHz in the potentiostatic mode with a disturbing sinusoidal signal at the potential amplitude of 10 mV over the corrosion potential. Then, after a few minutes or a longer interval when changes in the corrosion potential did not exceed the level of 0.1 mV/s, LPR tests were conducted in the potentiodynamic mode. The reinforcement was polarized at a rate of 1 mV/s within the range of potential changes from −150 mV to +50 mV regarding the corrosion potential.

The first measuring series began two weeks after placing specimens in concrete. Other five measuring series were conducted in one- and two-week intervals. At the next stage, six measuring series were taken at a two-week interval. Final measuring series took place in a month interval. After 18 months, prior to scheduled microscopic measurements, EIS and LPR tests were repeated but only on four specimens representing each of *B, G, B-Cl,* and *G-Cl* groups. At the final testing stage, results obtained from polarization measurements were compared with microscopic observations and analyses.

3.2. Structural Tests

The specimens for structural tests were cut out perpendicularly to the rebar axis. Metallographic specimens were prepared from those cut pieces of concrete with rebars made of steel B500SP. Tests on the macrostructure of concrete specimens were performed at the stand specially prepared for macro testing. And microstructure of concrete was tested using Olympus SZX9 stereo microscope and GX51 inverted microscope (LM) (Olympus Corporation, Tokyo, Japan).

Advanced tests on the specimen structure at selected points were performed with HITAHI S-3400N scanning electron microscope (SEM) (Hitachi, Tokyo, Japan) at an accelerating voltage within the range of 15–25 kV. Tests on chemical composition at selected micro-areas of the specimens were performed with the energy dispersive spectroscopy (EDS) using the spectrometer by THERMO NORAN company with SYSTEM SIX software (Thermo Fisher Scientific Inc., Waltham, MA, USA). As microscopic tests are invading into the structure of the specimen material, they were performed after the final stage of electrochemical tests. Therefore, those tests should be only related to results obtained after the last, 792 day of measurements.

4. Results from Electrochemical Tests and Their Analysis

4.1. Measuring Corrosion Potential of Reinforcement in Concrete

Figure 3 shows a comparison of results from 160 measurements of corrosion potential of the steel reinforcement in concrete from twelve test elements. Figure 3a presents results for concrete without any additives, whereas Figure 3b refers to concrete with calcium chloride. All measurements were performed against the silver chloride electrode.

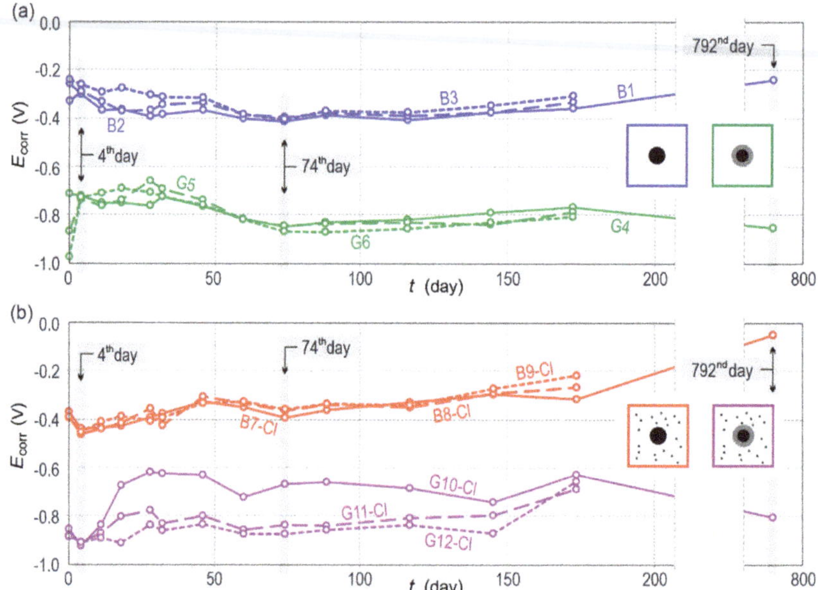

Figure 3. Changes in the function of time for corrosion potential E_{corr} of galvanized and black steel reinforcement (**a**) in concrete without chlorides, (**b**) in concrete with chlorides; selected measuring series presented as diagrams and analyzed in details as part of electrochemical impedance spectroscopy (EIS) and linear polarization resistance (LPR) measurements are marked with grey color.

Core tests were conducted in less than six months and restarted after about an 18-month interval, but only on four selected specimens that were further subjected to microscopic tests.

4.2. Linear Polarization Resistance (LPR) Tests

Linear polarization resistance (LPR) tests were the last tests from the group of electrochemical tests performed on the reinforced concrete specimen. Contrary to electrochemical impedance spectroscopy (EIS) tests regarded as complementary measurements, LPR tests were conducted on each test element from each measuring series. The whole period of testing produced a total of 160 polarization curves, in which exemplary shapes for three selected measuring series are illustrated in Figure 4. Characteristic time points for measurements were the following measuring days: 4, 74, and 792. The selected measuring days, marked with grey color in Figure 3, were to provide a more detailed presentation and analysis of test results from the initial phase of measurements (day 4), the halfway point of tests (day 74) and from the period directly before planned microscopic tests and analyses (day 792).

On day 4 of measurements, shapes of polarization curves (Figure 4a) were more arranged showing the clear formation of four groups. The corrosion potential of the specimens with black steel was less negative than in case of galvanized steel. The mean potential value in the specimen groups *B* and *B-Cl* was −0.28 V and −0.45 V, respectively. For the groups *G* and *G-Cl*, it was equal to −0.73 V and −0.91 V, respectively. However, the most significant information obtained from the analysis of polarization curves was the polarization resistance R_p was inversely proportional to the density of corrosion current i_{corr}. The Stern-Geary Equation (1) was used to calculate values of i_{corr}, taking into account coefficients of rectilinear slope for segments of polarization curves—anodic b_a and cathodic b_c.

$$i_{corr} = \frac{B}{R_p \cdot A}, \quad B = \frac{b_a b_c}{2.303 \cdot (b_a + b_c)} \tag{1}$$

Parameters of the Equation (1) taken for calculations of i_{corr} are shown in Appendix A—Table A1, which contains results for only 28 out of 160 analyzed polarization curves. Mean densities of corrosion current for individual groups used further in the paper, demonstrated that the lowest value of $i_{corr,mid}$ = 0.31 µA/cm^2 was obtained for the specimens from the group *B*. Because of naturally high alkaline pH (pH > 12.5) of concrete pore solution without additives, the obtained current density could indicate the passivation of reinforcing steel in concrete. Adding calcium chloride to the concrete mix caused the same black steel started to corrode in the specimens from the group *B-Cl* with $i_{corr,mid}$ = 1.65 µA/cm^2. And the zinc coating on the reinforcement also began to dissolve electrochemically at pH > 13.3 when in direct contact with highly alkaline pore solution. The mean density of corrosion current $i_{corr,mid}$ = 0.93 µA/cm^2 calculated for the group *G* could with high probability indicate that occurrence typical for the initial phase of concrete curing. The final, the highest mean density of corrosion current $i_{corr,mid}$ = 1.98 µA/cm^2 referred to galvanized reinforcing steel in concrete with chloride additives. For the specimen group *G-Cl*, it could indicate the accumulation of zinc corrosion caused by high pH of concrete pore solution and aggressive impact of chloride ions.

On day 74 of measurements, distributions of polarization curves (Figure 4b) were slightly shifted when compared to the distribution of the same specimens presented in Figure 4a. The corrosion potential of black steel in concrete without chlorides, averaged for the same group of specimens, was considerably reduced to $E_{corr,mid}$ = −0.40 V, and in concrete with chlorides the potential slightly increased to $E_{corr,mid}$ = −0.37 V. For galvanized steel in concrete without chlorides, a clear drop in potential to the value $E_{corr,mid}$ = −0.85 V was observed with the simultaneous increase in potential to the following level $E_{corr,mid}$ = −0.79 V. The comparison of graphical distributions of polarization curves from the beginning of tests (Figure 4a) and the halfway point of measurements (Figure 4b) can suggest that corrosion potential of black reinforcing steel had similar values as in case of galvanized steel, but with values at a greater spread.

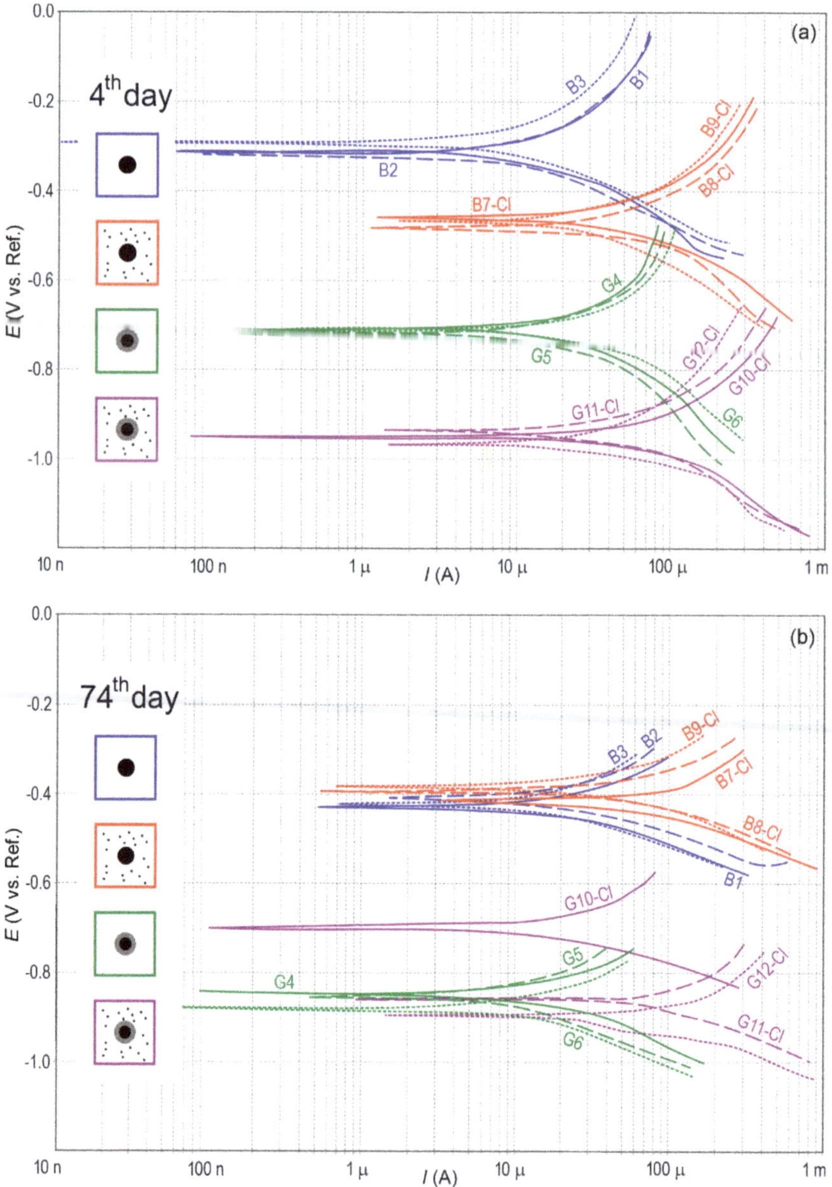

Figure 4. Potentiodynamic polarization curves for steel reinforcement in concrete from 12 specimens (**a**) on day 4 and (**b**) on day 74 of measurements: *B*—black steel in concrete without additives, *G*—galvanized steel in concrete without additives, *B-Cl*—black steel in concrete with chlorides, and *G-Cl*—galvanized steel in concrete with chlorides.

The analysis of calculated mean densities of corrosion current on day 74 of measurements indicated in concrete with chlorides still low values of $i_{corr.mid} = 0.27$ μA/cm^2 for black steel and a dramatic drop in $i_{corr.mid}$ to 0.14 μA/cm^2 for galvanized steel. The passivation of black steel in concrete without additives was a natural process, but no corrosion of zinc coating on steel reinforcement after several dozens of days in contact with the pore solution was desired despite the initially intensive dissolution of zinc. In concrete specimens with chlorides, the corrosion of black steel $i_{\text{corr.mid}} = 0.71\ \mu\text{A/cm}^2$ and galvanized steel $i_{\text{corr.mid}} = 1.16\ \mu\text{A/cm}^2$ was still observed on day 74 of measurements. But mean densities of corrosion current were slightly lower than on the first days of measurements.

After about two years of measurements (day 792), prior to the commencement of microscopic tests, polarization curves for individual specimens representing each of four groups (Figure 5) showed an interesting change. A noticeable change was found for black steel in concrete with chlorides (*B7-Cl*) because its corrosion potential largely increased to −0.04 V, but what is more important, the density of corrosion current dramatically decreased to just 0.01 $\mu\text{A/cm}^2$. Taking into account the fact the analyzed reinforced concrete specimen was potentially the most vulnerable to corrosion, the only possible explanation for the above situation was the high likelihood of the large accumulation of corrosion products (rust) around the steel rebar, and consequently, a serious limitation of direct access for oxygen and humidity to its surface. Hence, electrode processes on steel were stopped. For galvanized reinforcing steel in concrete with chlorides (*G10-Cl*), the corrosion current density dropped to 0.34 $\mu\text{A/cm}^2$. The likely reason was the accumulation of corrosion products of zinc on the coating slightly slowing down its further dissolution. After two years of measurements, black steel (*B1*) and galvanized steel (*G4*) in the concrete specimens without chlorides had stable but low density of corrosion current (0.35 $\mu\text{A/cm}^2$ and 0.29 $\mu\text{A/cm}^2$, respectively), which suggested the reinforcement corrosion was not likely to develop.

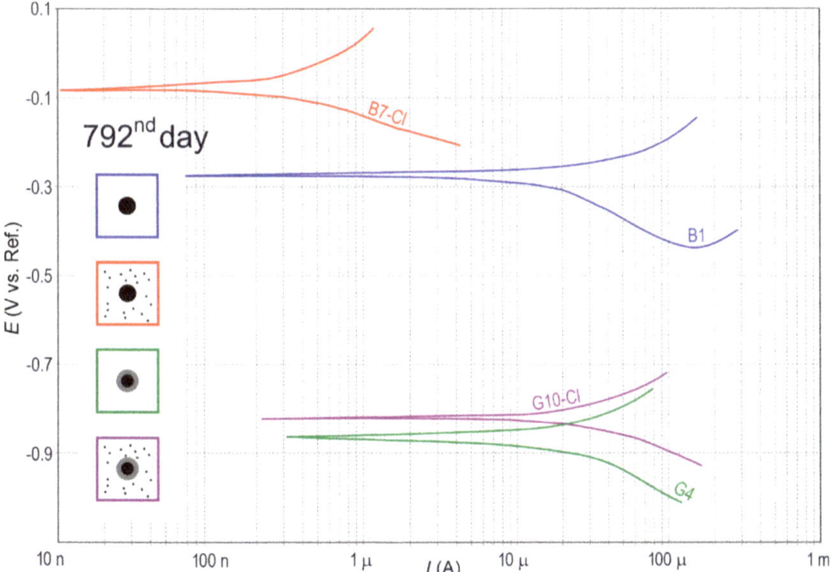

Figure 5. Potentiodynamic polarization curves for the steel reinforcement in concrete from four specimens on day 792, directly before microscopic tests: *B1*—black steel in concrete without additives, *G4*—galvanized steel in concrete without additives, *B7-Cl*—black steel in concrete with chlorides, and *G10-Cl*—galvanized steel in concrete with chlorides.

4.3. Electrochemical Impedance Spectroscopy (EIS) Tests

Electrochemical impedance spectroscopy (EIS) tests on the specimens of concrete with reinforcement were additional to core polarization tests performed by linear polarization resistance (LPR) method. Impedance tests focused only on three selected measuring series on day 4, 74, and 792.

Figure 6a illustrates in the complex plane results from impedance tests on the steel reinforcement in concrete from 12 specimens on day 4 of measurements, whereas Figure 6b presents the same results as the function of phase shift angle ϕ against the logarithm of measurement frequency f. For the specimens from groups B and G made of concrete without chlorides, spectra at the Nyquist plots (Figure 6a) exhibited capacitive behavior similar in shape to parts of semi-circles, in which extrapolation to the horizontal axis of real impedance provided the possibility of estimating charge transfer resistance R_t across the interface. It is interesting to note that semi-circles referring to galvanized steel had slightly smaller conventional diameter. The probable reason was the dissolution of the zinc coating in highly alkaline (pH > 13.3) of concrete pore solution.

For the specimens from groups B-Cl and G-Cl made of concrete with chlorides, the size of spectra at the Nyquist plot was much smaller and the second time-constant was clearly noticeable. Shapes of those spectra were rather characteristic for the development of metal (Fe or Zn) corrosion in concrete. Observations made for the Nyquist plot (Figure 6a) were additionally confirmed by the Bode plot (Figure 6b). Low frequencies typical for metal in concrete without additives (groups B and G) displayed quite wide angles of phase shift ($-20°$–$-40°$), whereas the specimens of concrete with chlorides (groups B-Cl and G-Cl) reached the maximum level of $-15°$. The observed shapes of those spectra, which were very distinctive, quite explicitly indicated the corrosion of metal in the specimens from groups B-Cl and G-Cl. For galvanized steel (group G) the situation was slightly less favorable, which could suggest not so intensive processes of zinc dissolution.

The above qualitative analysis of obtained spectra was confirmed by the quantitative analysis. For that purpose, the equivalent electrical circuit illustrated in Figure 6(a1), was proposed for spectra of concrete specimens without chlorides. It was the classic Randles circuit, in which the double layer capacity was replaced with a constant phase element CPE_0. Parameter R_c is resistance of concrete, R_t is charge transfer resistance across the interface. The electric circuit shown in Figure 6(a2), which simulated the porous coating, was used for concrete specimens with chlorides. Besides already mentioned parameters of the Randles circuit, that circuit had two characteristics of the electrochemical zinc coating on the steel reinforcement: R_f—resistance of the coating layer and CPE_f—constant phase element characterizing its quasi capacity. The Simplex method was used to find numerical values of elements in equivalent circuits, and those values are presented in Appendix A—Table A2. The summary table presenting calculated results, also contain numerical parameters of constant phase elements: CPE_0—Y_0 and α_0 and CPE_f—Y_f and α_0.

On day 74 of measurements, the impedance spectra at the Nyquist plot (Figure 7a) obtained for all four groups of the specimens were similar to spectra from day 4 (Figure 6a). The main difference referred to the specimens with galvanized steel, for which the spectra at the Nyquist plot had a slightly different shape resembling a typical image of metal passivated in concrete at low frequencies. That observation was confirmed by the Bode plot (Figure 7b), at which phase shift angles of galvanized steel at low frequencies were greater by ca. 5° (with reference to absolute values) than for black steel. Relationships between shapes of impedance spectra were rather typical for zinc behavior in hardened concrete without any additives.

For the specimens of concrete with chlorides, day 74 of measurements revealed a small drop in the intensity of corrosion processes of galvanized and black steel. The presented deduction resulted from the quantitative analysis on the basis of the equivalent electrical circuits illustrated in Figure 7(a1,a2), similar to diagrams shown in Figure 6(a1,a2). That phenomenon could be explained by a gradual accumulation of corrosion products of iron and zinc on the surface of steel reinforcement in the specimen, which additionally were deposited in concrete pores. Dissolution of metals (Fe and Zn) was slightly inhibited mainly due to the limited access of oxygen to the surface of both metals.

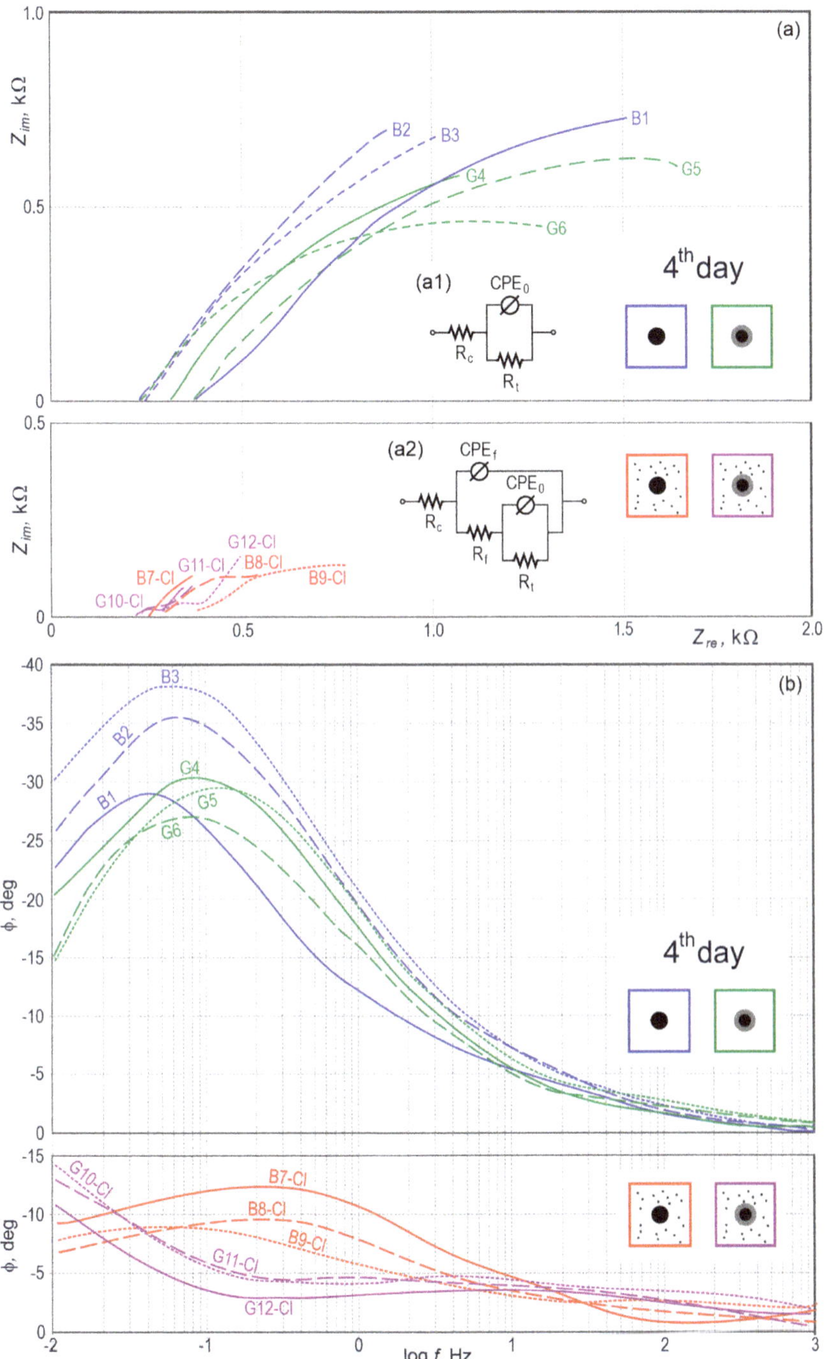

Figure 6. Results from impedance testing of steel reinforcement in concrete from 12 specimens on day 4 of measurements illustrated at the Nyquist (**a**) and Bode (**b**) plots, *B*—black steel in concrete without additives, *G*—galvanized steel in concrete without additives, *B-Cl*—black steel in concrete with chlorides, *G-Cl*—galvanized steel in concrete with chlorides; (**a1**) electrical equivalent circuit for the *B* and *G* spectrum groups, (**a2**) electrical equivalent circuit for the *B-Cl* and *G-Cl* spectrum groups.

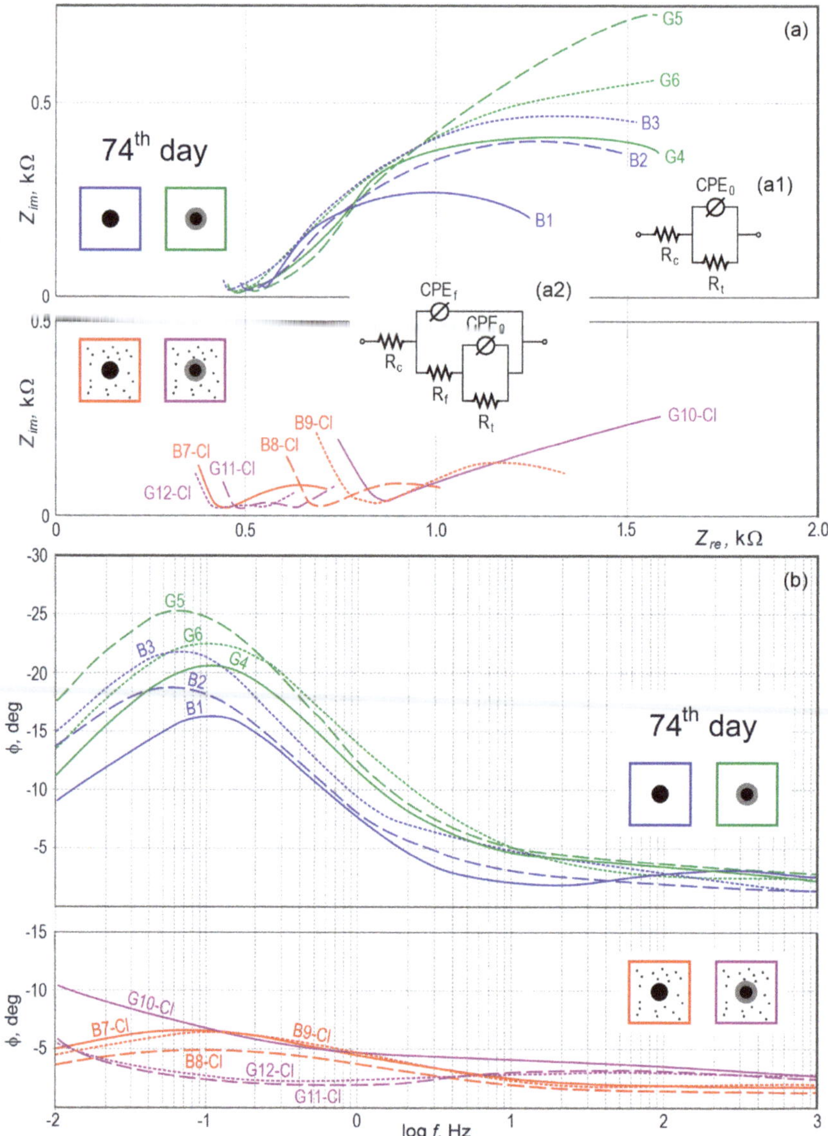

Figure 7. Results from impedance testing of steel reinforcement in concrete from 12 specimens on day 74 of measurements illustrated at the Nyquist (**a**) and Bode (**b**) plots, *B*—black steel in concrete without additives, *G*—galvanized steel in concrete without additives, *B-Cl*—black steel in concrete with chlorides, and *G-Cl*—galvanized steel in concrete with chlorides; (**a1**) electrical equivalent circuit for the *B* and *G* spectrum groups, (**a2**) electrical equivalent circuit for the *B-Cl* and *G-Cl* spectrum groups.

The last series of impedance measurement was made on day 792, that is, after about two years of measurements and included only four specimens (*B1*, *G4*, *B7-Cl*, *G10-Cl*) being the one-element representation of all four groups (*B*, *G*, *B-Cl*, *G-Cl*) of test elements. The Nyquist plot (Figure 8a) shows one dominating spectrum (*B7-Cl*) and three considerably smaller spectra near the beginning of the coordinate system. For better clearness of the results, those three spectra are again presented below, in the second complex plane with 20-time smaller values of complex and real impedance. The clear dominance of one spectrum over others indicated the impedance of the whole steel-concrete system in the specimen *B7-Cl* was about 20 times greater than the impedance of the other three specimens. As all impedance measurements on the final measuring series (day 792) were conducted within a broader range of frequencies 0.01 Hz–100 kHz, particularly the spectrum *B7-Cl* had a rather interesting shape. It had (from the left side) a high-frequency, slightly oblate semi-circle expressing electrochemical properties of concrete, and from the point of contraflexure a rectilinear part changing into the arc shape. The last described part could indicate diffusion control over the corrosion process. The spectrum shape for *B7-Cl* in the complex plane and at the Bode plot demonstrated a very strong slowdown on day 792 of measurements, comparable to the situation, in which dissolution of reinforcing steel in concrete was inhibited despite the presence of aggressive chloride ions. A continuous increase in the volume of corrosion products was supposed to inhibit seriously the access of oxygen to the steel surface, and consequently to eliminate the agent required for the development of electrochemical corrosion of steel in concrete.

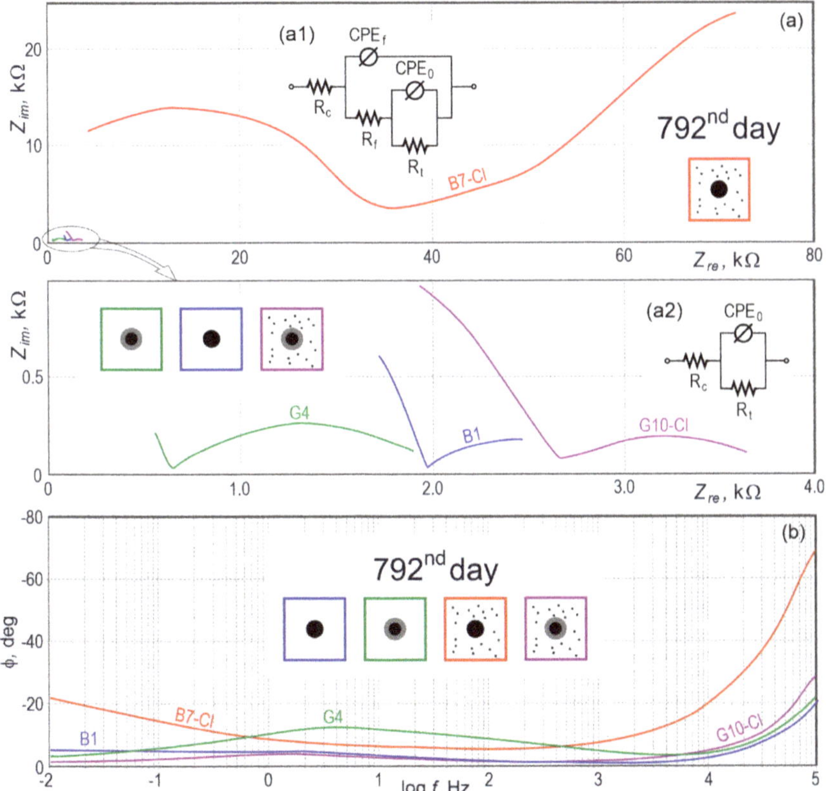

Figure 8. Results from impedance testing of steel reinforcement in concrete from 12 specimens on day 792 of measurements illustrated at the Nyquist (**a**) and Bode (**b**) plots, directly prior to microscopic tests *B1*—black steel in concrete without additives, *G4*—galvanized steel in concrete without additives, *B7-Cl*—black steel in concrete with chlorides, and *G10-Cl*—galvanized steel in concrete with chlorides; (**a1**) electrical equivalent circuit for the *B-Cl* spectra group, (**a2**) electrical equivalent circuit for the *B, G* and *G-Cl* spectrum groups.

The qualitative evaluation and adjustment with the Simplex method of the theoretical spectrum, according to the diagram shown in Figure 8(a1) confirmed the above observations as the determined charge transfer resistance R_t reached 1680 kΩcm^2, which corresponded to the very low density of corrosion current $i_{corr} = 0.01$ μA/cm^2

Impedance spectra of the other three specimens (*B1, G4, G10-Cl*) at the Nyquist plot were rather similar in shape, particularly with reference to the low-frequency range describing the properties of metal in concrete. The qualitative analysis conducted with the electrical equivalent circuit illustrated in Figure 8(a2) also confirmed similar values of charge transfer resistance $R_t = 34$ k–50 kΩcm^2, which corresponded to very low densities of corrosion current $i_{corr} = 0.29$–0.35 μA/cm^2. According to criteria published in [51], determined for rebars in concrete, such low densities of corrosion current usually indicate a very low level of steel corrosion. Moreover, each spectrum had a noticeably different inflexion point (that is, the closest point to the horizontal axis of real impedance), from which electrochemical properties of concrete were revealed within the low-frequency range. Diversified inflection points of those spectra indicated different values of concrete resistivity during measurements under the impact of alternating current flowing between reinforcing steel and the auxiliary electrode, placed on the top surface of the concrete specimen. The analyzed cases showed concrete in the specimen *G4* had the lowest resistivity, and concrete in the specimen *G10-Cl* had the highest resistivity.

5. Results from Structural Tests and Their Analysis

Corrosion tests indicated the change in parameters of corrosion current over time which could suggest structural changes at the concrete-rebar interface. Therefore, microstructure and chemical compositions at the interface between concrete and rebar made of steel B500SP were analyzed for all four tested cases after completing corrosion tests:

- black reinforcing steel B500SP in concrete without any additives,
- galvanized reinforcing steel B500SP in concrete without chlorides,
- black reinforcing steel B500SP in concrete with chlorides
- galvanized reinforcing steel B500SP in concrete with chlorides.

5.1. General Description of the Microstructure of Concrete Reinforced with Rebars Made of Steel B500SP (Stereo, LM)

Tests on concrete-reinforcement joints began from analyzing its microstructure after completing corrosion tests, using the stereo microscope (Figure 9). Figure 9a presents the cross-section (with dimensions of 40 mm × 40 mm) of the tested concrete specimen with a galvanized rebar concreted in the central point, made of ribbed steel of grade B500SP and with a diameter of ϕ8 mm.

Figure 9b illustrates the close-up of the cross-section of steel rebar with a clearly visible, thin, silver and circumferential zinc coating. There are noticeable particles of coarse and fine (sand) aggregates near the rebar. Aggregates are tied together with cement gel. Figure 9c,d show another close-up of a strongly diversified thickness of the zinc coating, whose thickest layer is observed near oblique ribs of the reinforcing steel.

The tests conducted on the microstructure of reinforced concrete specimens using the light microscope revealed good conditions of joints between concrete and the reinforcement for rebars made of black steel (Figure 10a). And simultaneous observations confirmed previous conclusions made on the basis of tests using the stereo microscope (Figure 9) concerning the non-uniform thickness of zinc coating on rebars (Figure 10b).

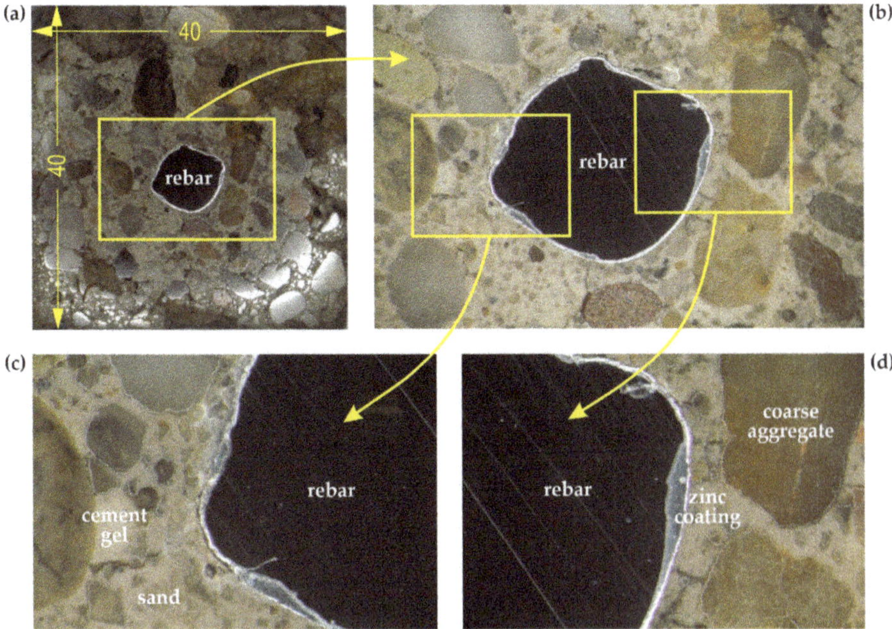

Figure 9. Cross-section of the concrete specimen with dimensions of 40 mm × 40 mm, containing galvanized rebar (stereoscope): (**a**) general view of a cross-section of a steel bar in concrete (**b**) detail marked with a rectangle from Figure 9a, (**c**,**d**) detail marked with a rectangle from Figure 9b.

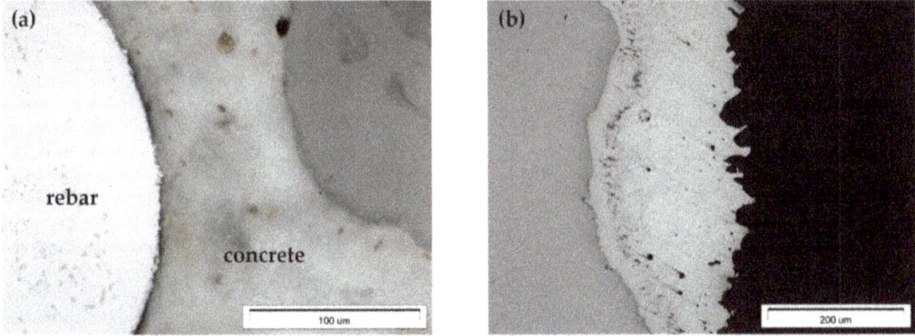

Figure 10. Microstructure of the joint between concrete and rebar made of galvanized steel B500SP (LM): (**a**) general view of a cross-section of a steel bar in concrete (**b**) morphology of the zinc coating on the rebar in concrete.

5.2. Evaluation of Microstructure and Chemical Compositions of Joints Between Concrete and Rebar (SEM, EDS)

5.2.1. Black Reinforcing Steel B500SP in Concrete with and without Chlorides

The analysis of the microstructure of joints between concrete without chlorides and rebar made of B500SP black steel (Figure 11) showed a difference in concrete structure regarding its phase composition (Figure 11a—image from BSE detector) and rebar correctly embedded in concrete. The analysis of chemical composition at the interface between concrete and rebar (Figure 12) indicated the presence of elements being components of concrete or steel rebar. Oxygen (Figure 12b), probably in the form of oxide, was found on the rebar surface. Oxidation of zinc coating is a characteristic effect of its destruction [52].

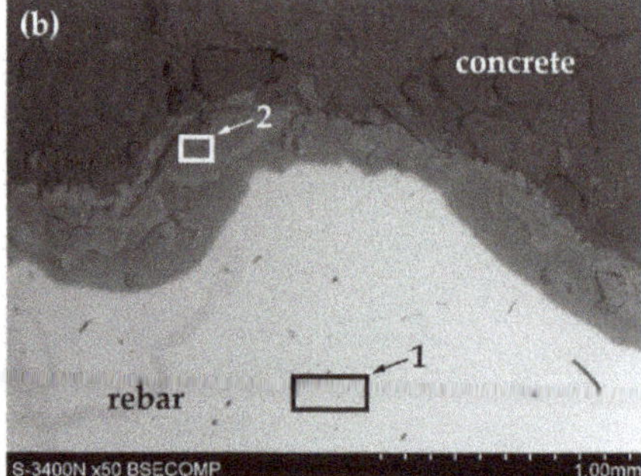

Figure 11. Microstructure of the joint between concrete without chlorides and reinforcement made of black steel B500SP (SEM): (**a**) general view of a cross-section of a steel bar in concrete (**b**) morphology of the boundary of separation rebar–concrete with marked places 1 and 2 for microanalysis of the chemical composition.

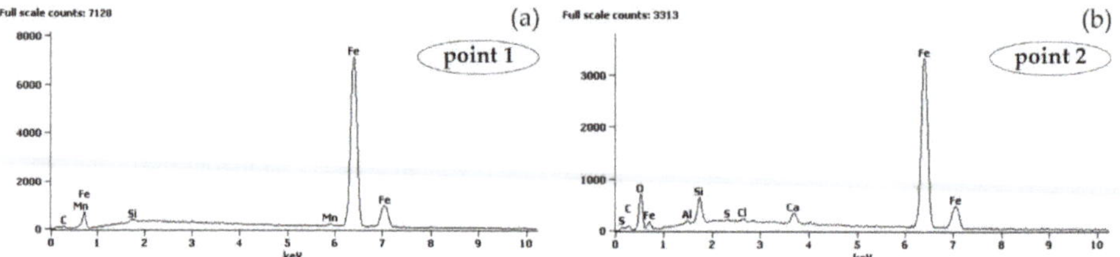

Figure 12. Spectra obtained from energy-dispersive X-ray spectroscopy (EDS) method for areas marked in Figure 11b: (**a**) from point 1 (**b**) from point 2.

As expected, no traces of corrosion damage at the interface between steel and concrete were found, what was normal in the highly alkaline environment of concrete. Results from polarization tests on corrosion current density also indirectly suggested the passivation of steel, which was typical for such a system.

The analysis of joints between concrete with chlorides and black reinforcing steel B500SP indicated a clear change in the nature of the joint when compared to concrete without chlorides (Figure 11). The presence of chlorides in concrete resulted in a large number of corrosion products at the rebar/concrete interface (Figure 13). Corrosion products of the steel reinforcement (rust) did not only accumulate on the surface of the rebar, but also penetrated the porous structure of concrete. A considerable concentration of Cl and oxygen (Figure 14b) was found in these products which probably indicated the presence of chlorides and oxides in that area.

No concrete cracking in the interfacial zone was seen which indicated critical stresses blowing out the concrete cover from the inside did not occur. It should be noticed that the layer of corrosion products was thick enough to make the diffusion of oxygen to the steel surface very difficult. That effect was evident in results from impedance tests (Figure 8a) due to the appearance of the additional time constant.

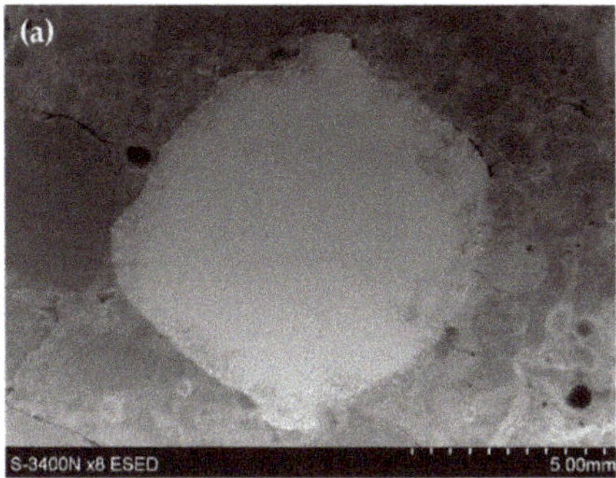 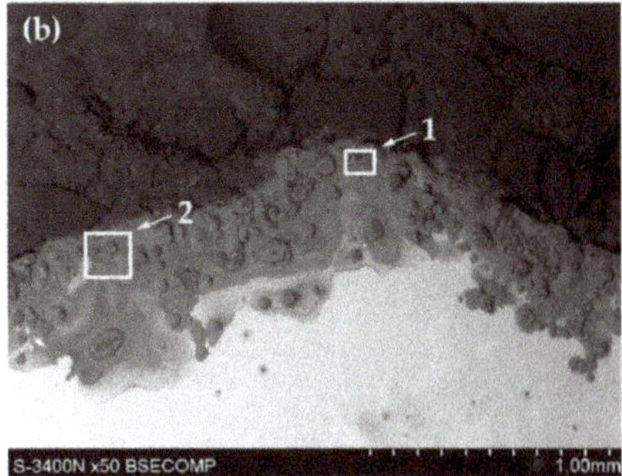

Figure 13. Microstructure of the joint between concrete with chlorides and black reinforcing steel B500SP (SEM): (**a**) general view of a cross-section of a steel bar in concrete (**b**) morphology of the boundary of separation rebar – concrete with marked places 1 and 2 for microanalysis of the chemical composition.

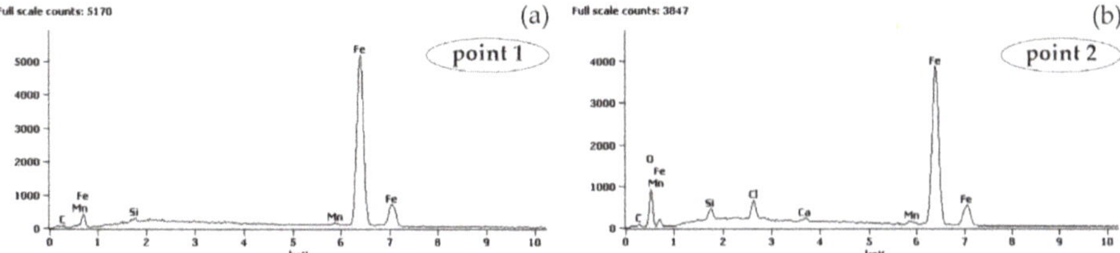

Figure 14. Spectra obtained from energy-dispersive X-ray spectroscopy (EDS) method for areas marked in Figure 13b: (**a**) from point 1 (**b**) from point 2.

5.2.2. Reinforcing Steel B500SP with Zinc Coating, in Concrete with and without Chlorides

The analysis of the joint between concrete without chlorides and the reinforcing steel B500SP with zinc coating demonstrated a clear difference the coating thickness (Figure 15). It varied from more than ten (78.4 μm) to over 600 μm, but usually was within the range from 150 μm to 250 μm. The thickest zinc coating (620 μm) was near one of oblique ribs of the rebar. The variable thickness of the zinc coating is an effect directly resulting from the technology of the galvanizing process, but it can also be the result of the preparation of metallographic samples—the zinc "rubs" on the surface of the specimen. In this case, it is the effect of galvanizing technology. The analysis of chemical composition for the section of the zinc coating indicated its great uniformity (Figure 16). That result was confirmed by testing the relative concentration elements on the surface (mapping) of the concrete-rebar joint (Figure 17).

Polarization tests on the galvanized reinforcement in concrete without chlorides indicated some increase in corrosion current density on first days of measurements when compared to black steel. This effect is usually noticed during the first weeks of concrete curing. In this case, the microscopic analysis referred to a two-year old concrete, for which processes of dissolution of zinc coating on steel are not usually observed or have been stabilized. Therefore, it is difficult to determine the original thickness of the zinc coating. It may be wondered whether it's very non-uniform thickness observed under the microscope was only caused by the specificity of the technological process of the formation of hot-dip galvanized zinc coating or the effect of partial dissolution of zinc in the highly alkaline environment of concrete pore solution.

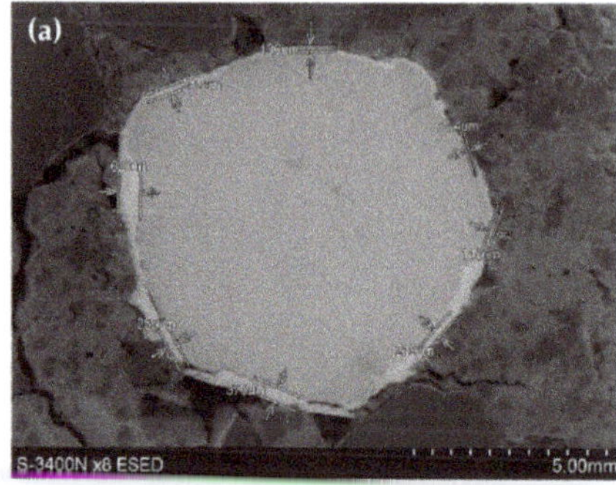
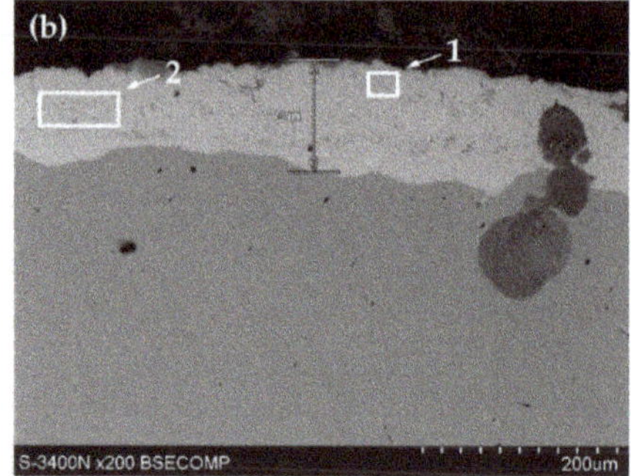

Figure 15. Microstructure of the joint between concrete without chlorides and reinforcing steel B500SP with zinc coating (SEM): (**a**) general view of a cross-section of a steel bar in concrete (**b**) morphology of the boundary of separation rebar - concrete with marked places 1 and 2 for microanalysis of the chemical composition.

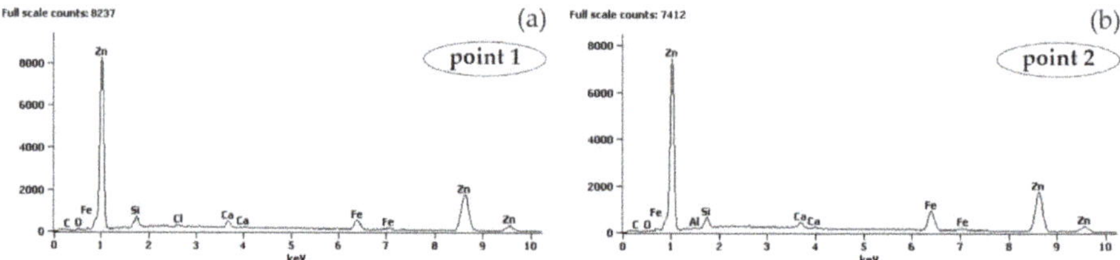

Figure 16. Spectra obtained from energy-dispersive X-ray spectroscopy (EDS) method for areas marked in Figure 15b: (**a**) from point 1 (**b**) from point 2.

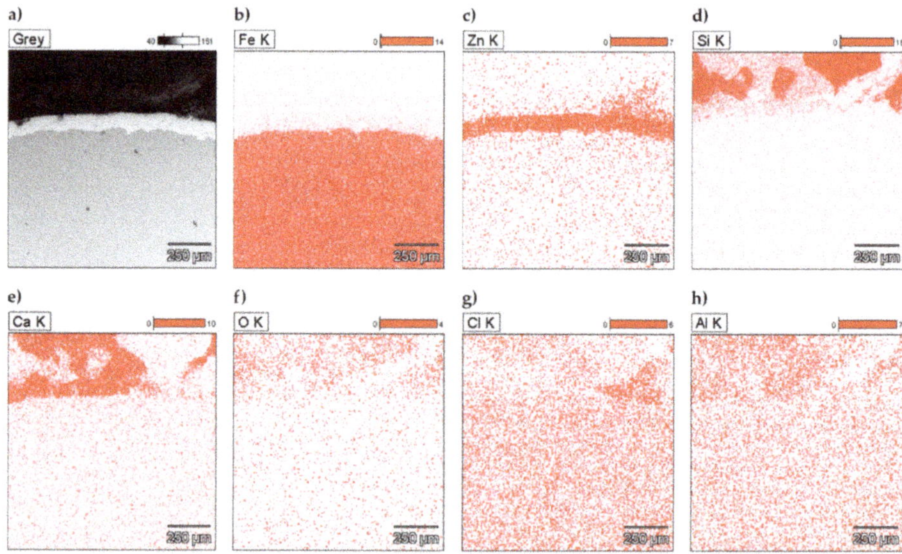

Figure 17. (**a**) Microstructure of the surface of tested concrete without chlorides—reinforcing steel B500SP with zinc coating (SEM) and relative concentrations of elements on the surface (mapping): (**b**) Fe, (**c**) Zn, (**d**) Si, (**e**) Ca, (**f**) O, (**g**) Cl, and (**h**) Al (EDS).

The analysis of joint between concrete with chlorides and galvanized reinforcing steel B500SP indicated significant non-uniformity of the zinc coating on the rebar with reference to its thickness and morphology (Figure 18), and a similar situation was observed for the bond when no chlorides were present.

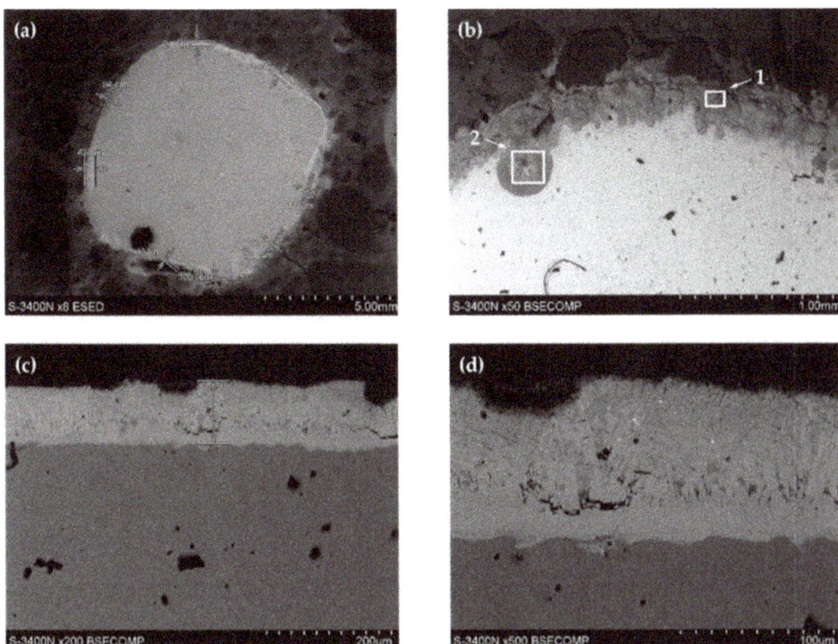

Figure 18. Microstructure of the joint between concrete without chlorides and galvanized reinforcing steel B500SP (SEM): (**a**) general view of a cross-section of a steel bar in concrete (**b**) morphology of the boundary of separation rebar–concrete with marked places 1 and 2 for microanalysis of the chemical composition (**c**) morphology of the zinc coating on the rebar in concrete (**d**) area of the zinc coating with the crack shown in Figure 18c.

The thickness of the zinc coating varied from as low as 100 μm (94.4 μm) to over 400 μm (434 μm) (Figure 18a). Morphology of the coating across its cross-section (Figure 18d) suggested a diverse growth of crystallites in individual micro-areas. The analysis of chemical composition across the coating section (Figures 19 and 20) demonstrated considerable changes, particularly for concentration of Cl, which could locally reach high values (Figure 20b). Chlorine was also the element, which occurred along with the zinc coating (Figure 20c,g).

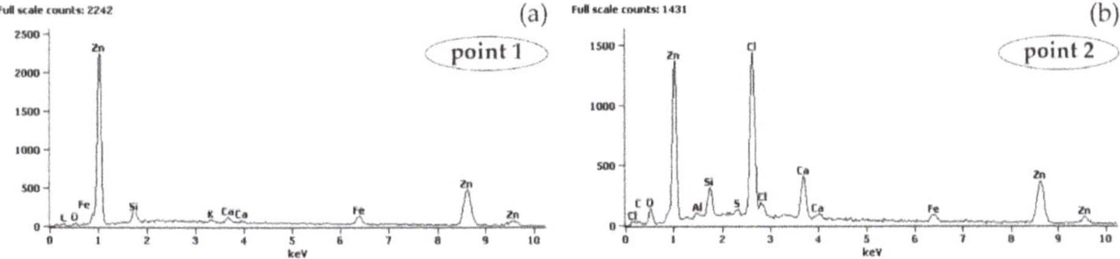

Figure 19. Spectra obtained from energy-dispersive X-ray spectroscopy (EDS) method for areas marked in Figure 18b: (**a**) from point 1 (**b**) from point 2.

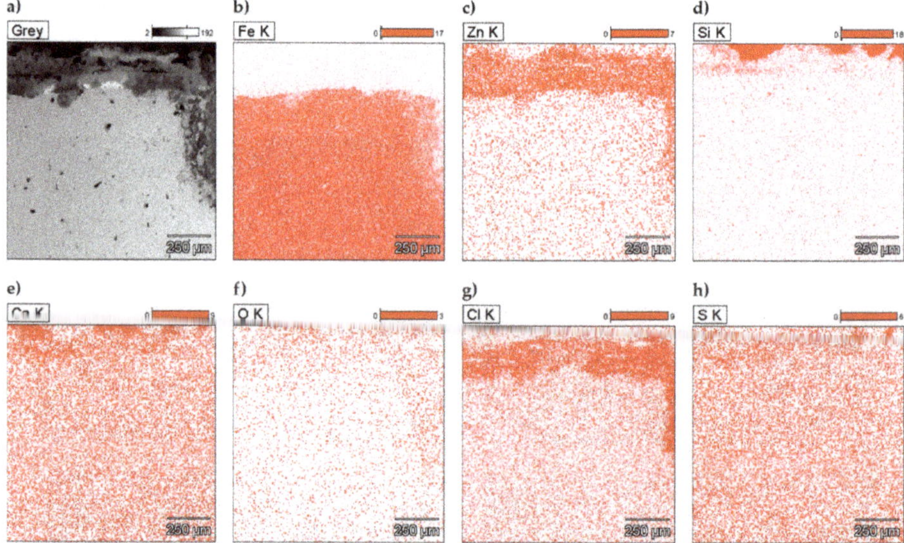

Figure 20. (a) Microstructure of the surface of tested concrete without chlorides—reinforcing steel B500SP with zinc coating (SEM) and relative concentrations of elements on the surface (mapping): (b) Fe, (c) Zn, (d) Si, (e) Ca, (f) O, (g) Cl, and (h) Al (EDS).

The comparison of the thickness of zinc coating on steel in concrete without chlorides (Figure 15) and with chlorides (Figure 18) (though they are different specimens) indicated the presence of chlorides significantly reduced its thickness. That phenomenon was expected as a high concentration of chlorides always causes the corrosion of zinc coating [33]. The analysis of microscopic images showed the zinc coating at any place was completely damaged. Hence, high densities of corrosion current recorded during tests were always related to the dissolution of zinc, and not reinforcing steel protected by zinc. Corrosion products of zinc, with a smaller volume than those of iron, also did not cause cracks in the concrete cover of the reinforcement. Apparently, corrosion products of zinc filling the porous structure of concrete in the interfacial zone had enough space, and no undesirable critical values of tensile stress in concrete were triggered.

6. Discussion

The analysis of changes in a function of time of corrosion potential E_{corr} of steel reinforcement in concrete very clearly demonstrated (Figure 4) that corrosion potential of black reinforcing steel, both in case of concrete without additives (group *B*), and with chlorides (group *B-Cl*) was relatively close within the range of −0.1––0.5 V. While evaluating obtained results with reference to standard criteria [53] for testing potential of the reinforcement in reinforced concrete structures, the information about the corrosion probability of 50%–95% was obtained after calculating results per the potential of copper-sulphate electrode. It should be pointed out that the lowest values of potential up to the level of −0.45 V were observed for concrete with chlorides during the first weeks of testing. And distributions of corrosion potential of galvanized reinforcing steel in line with expectations were characterized by lower values which directly resulted from zinc position in the electrochemical series of metals. In concrete without chlorides, zinc as less noble metal than iron, being the main component of tested low-carbon reinforcing steel, had the potential of ca. 0.3–0.5 V lower than the potential of black steel when in direct contact with concrete pore solution. For concrete with chlorides, the measured corrosion potential of galvanized reinforcement demonstrated the similar difference in potentials. However, the spread of potential values for the specimens from the group *G-Cl* was considerably greater than for the group *B-Cl*.

From a scientific perspective, the analysis of results from electrochemical tests (particularly density of corrosion current) seems to be very interesting during a 2-year period of performed tests. Figure 21 presents the comparison of changes in corrosion current densities i_{corr} within two years of measurements, obtained on the basis of analyses of polarization curves of reinforcing steel in the concrete specimens. Results for galvanized steel (*G*) and black steel (*B*) in concrete without additives are shown in Figure 21a, whereas results for similar rebars in concrete with chlorides are illustrated in Figure 21b.

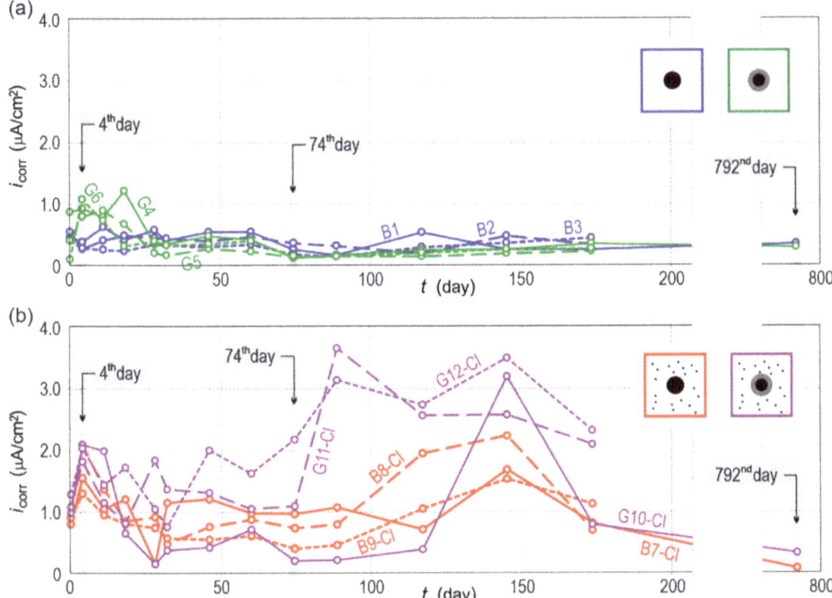

Figure 21. Changes in the function of time for corrosion current densities i_{corr} of galvanized and black steel reinforcement (**a**) in concrete without chlorides, (**b**) in concrete with chlorides; selected measuring series presented as diagrams and analyzed in details as part of EIS and LPR measurements are marked with grey color.

As shown in Figure 21a, black reinforcing steel in concrete without additives (*B*) generally had stable but the low density of corrosion current through the whole period of measurements, which in accordance with criteria for evaluating corrosion risk to steel in concrete, could indicate the passivation. Microscopic tests (Figure 12b) demonstrated the presence of oxygen on the surface of steel rebar which probably indicated the presence of a passive layer of oxides that is characteristic for highly alkaline concrete pore solution (pH > 12.5).

In case of galvanized steel reinforcement in concrete without chlorides (*G*), elevated values of corrosion current density (oscillating around 1 µA/cm²) were observed in the initial phase of measurements (until day 18 of measurements). Then, corrosion current density stabilized at the level of 0.3–0.5 µA/cm², which was a slightly better result than for black steel. The already mentioned quite a rapid increase in corrosion rate for galvanized steel in concrete without chloride content was related to zinc corrosion in highly alkaline concrete pore solution. At pH > 13.3 corrosion of galvanized coat on reinforcement was initiated.

Considering galvanized and black steel in concrete with chlorides, dynamics of changes in corrosion current densities and spread of those values during two-year measurements was very diversified (Figure 21b). During first 74 days of measurements, density of corrosion current for black steel did not exceed the level of 1.3 μA/cm^2, whereas for galvanized steel values of i_{corr} were higher, even up to 2 μA/cm^2. After day 74, a further increase in corrosion current densities was observed for both galvanized and black steel. However, a wider spread of values was found for reinforcing steel protected with the coating. Within that period, values of corrosion current density of the specimens with galvanized steel reached very high values of 3.8 μA/cm^2. According to microscopic tests, zinc was found to be the only corrodible metal (Figure 20) on the undamaged structure of steel matrix which, considering the operational safety of reinforced concrete structures, is very important and positive information.

To sum it up, there was very strong corrosion of the reinforcement within the whole discussed period of two years of electrochemical testing according to criteria [51] for evaluating corrosion risk to reinforcing steel. Interestingly, after an 18-month interval in measurements determined densities of corrosion current dramatically decreased on day 792 of measurements. Corrosion of black steel was just inhibited as an insulating barrier from corrosion products of iron was formed. The microscopic tests (Figure 13) showed that a layer of corrosion products was thick enough to hinder diffusion of oxygen required for corrosion development. Diffusion-controlled electrochemical processes were also confirmed by the Nyquist plot (Figure 8a). It should be pointed out that the specimens were kept under rather stable thermal and humid conditions (temperature 20 ± 2 °C and relative humidity 50 ± 10%) during a very long, 18-month interval in measurements.

It is worth mentioning that despite an intensive development of corrosion on black steel in concrete with chlorides, no cracks were found on concrete surface of the test elements. It could be explained by a small diameter of rebars with reference to the thickness of the concrete cover. Thus, the volume of accumulating corrosion products did not exceed the available volume of concrete pores adjacent to rebars. Consequently, no critical values of tensile stress in concrete were observed.

7. Conclusions

The following conclusions can be drawn from two-year comparative tests on efficiency of the hot-dip galvanized zinc coating applied on reinforcing steel B500SP in concrete.

- Comparative tests of changes in the density of corrosion current in the concrete specimens without chlorides basically demonstrated no development of corrosion, and possible passivation was expected in case of black steel. Higher densities of corrosion current were observed for galvanized steel during the first days of testing. The reason was the dissolution of zinc after the contact with initially high pH of concrete pore solution.
- Six-month measurements demonstrated a higher density of corrosion current in the concrete specimens with high 3% concentration of chlorides, which unambiguously indicated corrosion in concrete reinforced with galvanized or black steel. Microscopic observations indicated zinc was the only metal corroding with the reinforcement made of galvanized steel, while the rebar surface showed no signs of corrosion.
- Densities of corrosion current determined for selected specimens dramatically decreased after an 18-month interval in measurements. Corrosion was inhibited on black steel as an insulating barrier of corrosion products was formed. That result was also confirmed with microscopic observations.
- The analysis of structural bonds in steel without or with zinc coating, containing concrete without or with chlorides demonstrated that the coating formed during hot dip galvanizing had varied thickness measured in the circumference at the cross-section of the rebar where the thickest layer of zinc was always observed at oblique ribs of the rebar.

- Additionally, the structural tests showed that the chemical composition at the section of the zinc coating was very homogeneous, and the layer of corrosion products indicated an increased content of oxygen, particularly for zinc coating in concrete with chlorides.
- Results obtained from corrosion (LPR, EIS) and structural (SEM, EDS) tests on the specimens of concrete reinforced with steel B500SP demonstrated a very favorable impact of zinc coating on rebars by providing effective protection against corrosion in chloride environment.

Author Contributions: M.J.: conceptualization, methodology, analysis and description of electrochemical measurements, writing original draft; M.S.: methodology of microscopic examinations, their analysis and description, supervision over the whole paper; J.K.: preparation of concrete specimens with reinforcement, performance of electrochemical measurements; B.C.: performance of microscopic examinations and preparation of test results. All authors have read and agreed to the published version of the manuscript.

Funding: The research was financed by Silesian University of Technology (Poland) within the grant no. 03/020/RGP_19/0072, and in part within the project BK-200/RM0/2020.

Conflicts of Interest: The authors declare no conflict of interest. The funders had no role in the design of the study; in the collection, analyses, or interpretation of data; in the writing of the manuscript, or in the decision to publish the results.

Appendix A

Table A1. Comparison of results from analyzing polarization curves of reinforcing steel in concrete, obtained for three selected measuring series on day 4, 74, and 792.

Specimen No.	Δt (Day)	$E_{Ag|AgCl}$ (V)	b_a (mV)	b_c (mV)	B (V)	R_p (kΩ)	R_pA (kΩcm^2)	i_{corr} (μA/cm^2)
B1	4	−0.300	176.8	122.5	0.03142	2.647	80.48	0.39
B2	4	−0.290	180.3	73.5	0.02267	2.719	82.65	0.27
B3	4	−0.261	187.4	82.6	0.02489	2.924	88.90	0.28
G4	4	−0.722	267.3	167.9	0.04477	1.598	48.58	0.92
G5	4	−0.735	159.0	205.9	0.03896	1.609	48.92	0.80
G6	4	−0.726	153.6	182.3	0.03620	1.108	33.67	1.08
B7-Cl	4	−0.437	167.5	128.0	0.03151	0.668	20.31	1.55
B8-Cl	4	−0.460	166.8	155.3	0.03492	0.552	16.78	2.08
B9-Cl	4	−0.452	161.9	115.4	0.02926	0.738	22.45	1.30
G10-Cl	4	−0.919	177.9	112.4	0.02991	0.472	14.34	2.09
G11-Cl	4	−0.905	177.6	103.1	0.02833	0.513	15.59	1.82
G12-Cl	4	−0.906	217.9	177.1	0.04242	0.690	20.97	2.02
B1	74	−0.411	28.8	49.8	0.00793	1.010	30.71	0.26
B2	74	−0.393	41.1	48.7	0.00969	0.866	26.33	0.37
B3	74	−0.404	42.6	49.4	0.00993	1.841	55.97	0.18
G4	74	−0.845	21.9	71.7	0.00728	2.001	60.84	0.12
G5	74	−0.844	40.1	63.2	0.01065	2.582	78.49	0.14
G6	74	−0.867	35.1	70.9	0.01020	2.161	65.69	0.16
B7-Cl	74	−0.390	50.7	52.8	0.01123	0.380	11.55	0.97
B8-Cl	74	−0.362	48.9	59.6	0.01166	0.517	15.73	0.74
B9-Cl	74	−0.357	39.4	58.4	0.01021	0.811	24.65	0.41
G10-Cl	74	−0.664	34.6	40.5	0.00810	1.209	36.75	0.22
G11-Cl	74	−0.835	176.9	84.7	0.02486	0.747	22.70	1.09
G12-Cl	74	−0.870	69.3	61.6	0.01417	0.215	6.54	2.17
B1	792	−0.240	77.9	68.2	0.01580	1.477	44.90	0.35
G4	792	−0.851	53.2	90.6	0.01456	1.646	50.04	0.29
B7-Cl	792	−0.040	58.6	45.1	0.01107	55.292	1680.87	0.01
G10-Cl	792	−0.806	51.9	56.0	0.01169	1.121	34.07	0.34

Table A2. Comparison of results from analyzing impedance spectra of reinforcing steel in concrete, obtained for three selected measuring series on day 4, 74, and 792.

| Specimen No. | Δt | $E_{Ag|AgCl}$ | R_c | R_f | CPE_f | | R_t | CPE_0 | | i_{corr} |
| | | | | | Y_f | α_f | | Y_0 | α_0 | |
	(Day)	(V)	(Ω)	(Ω)	($Fs^{\alpha-1}$)	-	(Ω)	($Fs^{\alpha-1}$)	-	($\mu A/cm^2$)
B1	4	−0.305	415	–	–	–	3388	2.31×10^{-3}	0.597	0.31
B2	4	−0.289	234	–	–	–	3634	2.16×10^{-3}	0.615	0.21
B3	4	−0.265	250	–	–	–	5925	1.98×10^{-3}	0.622	0.14
G4	4	−0.721	317	–	–	–	2181	1.89×10^{-3}	0.660	0.68
G5	4	−0.732	384	–	–	–	2324	1.76×10^{-3}	0.625	0.55
G6	4	−0.728	247	–	–	–	1618	2.02×10^{-3}	0.661	0.74
B7-Cl	4	−0.437	253	–	–	–	468	3.52×10^{-3}	0.539	2.21
B8-Cl	4	−0.458	294	–	–	–	376	3.98×10^{-3}	0.529	3.06
B9-Cl	4	−0.451	415	–	–	–	503	4.77×10^{-3}	0.528	1.91
G10-Cl	4	−0.915	216	75	2.12×10^{-3}	0.448	333	6.24×10^{-2}	0.641	2.96
G11-Cl	4	−0.902	228	95	2.71×10^{-3}	0.431	371	3.49×10^{-2}	0.643	2.51
G12-Cl	4	−0.905	276	138	1.90×10^{-3}	0.424	539	3.03×10^{-2}	0.648	2.59
B1	74	−0.406	548	–	–	–	857	2.72×10^{-3}	0.708	0.30
B2	74	−0.392	567	–	–	–	1388	2.63×10^{-3}	0.647	0.23
B3	74	−0.405	554	–	–	–	1565	2.40×10^{-3}	0.681	0.21
G4	74	−0.846	539	–	–	–	1430	1.86×10^{-3}	0.673	0.17
G5	74	−0.846	613	–	–	–	2863	1.48×10^{-3}	0.631	0.12
G6	74	−0.870	514	–	–	–	2097	1.39×10^{-3}	0.608	0.16
B7-Cl	74	−0.378	449	–	–	–	423	4.53×10^{-3}	0.399	0.87
B8-Cl	74	−0.349	708	–	–	–	417	3.50×10^{-3}	0.456	0.92
B9-Cl	74	−0.348	849	–	–	–	684	2.36×10^{-3}	0.456	0.49
G10-Cl	74	−0.663	890	382	6.14×10^{-4}	0.447	1007	6.52×10^{-3}	0.618	0.26
G11-Cl	74	−0.831	477	175	6.71×10^{-4}	0.426	465	5.32×10^{-2}	0.637	1.76
G12-Cl	74	−0.865	419	151	8.49×10^{-4}	0.441	488	5.57×10^{-2}	0.612	0.95
B1	792	−0.199	1904	–	–	–	1262	7.69×10^{-4}	0.346	0.41
G4	792	−0.833	630	–	–	–	1413	2.19×10^{-4}	0.462	0.34
B7-Cl	792	−0.052	30,440	18,880	7.00×10^{-6}	0.420	735,900	6.54×10^{-5}	0.498	0.00
G10-Cl	792	−0.806	2620	–	–	–	1222	3.67×10^{-4}	0.421	0.31

References

1. Dong, B.; Fang, G.; Liu, Y.; Dong, P.; Zhang, J.; Xing, F.; Hong, S. Monitoring reinforcement corrosion and corrosion-induced cracking by X-ray microcomputed tomography method. *Cem. Concr. Res.* **2017**, *100*, 311–321. [CrossRef]
2. Angst, U.M.; Geiker, M.R.; Michel, A.; Gehlen, C.; Wong, H.; Isgor, O.B.; Elsener, B.; Hansson, C.M.; François, R.; Hornbostel, K.; et al. The steel–concrete interface. *Mater. Struct.* **2017**, *50*, 143. [CrossRef]
3. Fang, C.; Lundgren, K.; Chen, L.; Zhu, C. Corrosion influence on bond in reinforced concrete. *Cem. Concr. Res.* **2004**, *34*, 2159–2167. [CrossRef]
4. Mak, M.W.T.; Desnerck, P.; Lees, J.M. Corrosion-induced cracking and bond strength in reinforced concrete. *Constr. Build. Mater.* **2019**, *208*, 228–241. [CrossRef]
5. Imperatore, S.; Rinaldi, Z.; Drago, C. Degradation relationships for the mechanical properties of corroded steel rebars. *Constr. Build. Mater.* **2017**, *148*, 219–230. [CrossRef]
6. Duffó, G.S.; Farina, S.B. Electrochemical behaviour of steel in mortar and in simulated pore solutions: Analogies and differences. *Cem. Concr. Res.* **2016**, *88*, 211–216. [CrossRef]
7. Stefanoni, M.; Angst, U.; Elsener, B. Corrosion rate of carbon steel in carbonated concrete—A critical review. *Cem. Concr. Res.* **2018**, *103*, 35–48. [CrossRef]
8. Cheng, H.; Li, H.N.; Wang, D.S. Prediction for lateral deformation capacity of corroded reinforced concrete columns. *Struct. Des. Tall Spec. Build.* **2019**, *28*, 1–15. [CrossRef]
9. Ou, Y.C.; Susanto, Y.T.T.; Roh, H. Tensile behavior of naturally and artificially corroded steel bars. *Constr. Build. Mater.* **2016**, *103*, 93–104. [CrossRef]
10. Rocha, M.; Brühwiler, E.; Nussbaumer, A. Fatigue behaviour prediction of steel reinforcement bars using an adapted Navarro and de Los Rios model. *Int. J. Fatigue* **2015**, *75*, 198–204. [CrossRef]
11. Liu, X.; Niu, D.; Li, X.; Lv, Y.; Fu, Q. Pore solution pH for the corrosion initiation of rebars embedded in concrete under a long-term natural carbonation reaction. *Appl. Sci.* **2018**, *8*, 128. [CrossRef]

12. Cao, C. 3D simulation of localized steel corrosion in chloride contaminated reinforced concrete. *Constr. Build. Mater.* **2014**, *72*, 434–443. [CrossRef]

13. Ji, Y.S.; Zhao, W.; Zhou, M.; Ma, H.R.; Zeng, P. Corrosion current distribution of macrocell and microcell of steel bar in concrete exposed to chloride environments. *Constr. Build. Mater.* **2013**, *47*, 104–110. [CrossRef]

14. Zhang, Y.; DesRoches, R.; Tien, I. Impact of corrosion on risk assessment of shear-critical and short lap-spliced bridges. *Eng. Struct.* **2019**, *189*, 260–271. [CrossRef]

15. Moreno, J.D.; Bonilla, M.; Adam, J.M.; Borrachero, M.V.; Soriano, L. Determining corrosion levels in the reinforcement rebars of buildings in coastal areas. A case study in the Mediterranean coastline. *Constr. Build. Mater.* **2015**, *100*, 11–21. [CrossRef]

16. Papakonstantinou, K.G.; Shinozuka, M. Probabilistic model for steel corrosion in reinforced concrete structures of large dimensions considering crack effects. *Eng. Struct.* **2013**, *57*, 306–326. [CrossRef]

17. Sassine, E.; Laurens, S.; François, R.; Ringot, E. A critical discussion on rebar electrical continuity and usual interpretation thresholds in the field of half-cell potential measurements in steel reinforced concrete. *Mate. Struct./Mater. Constr.* **2018**, *51*. [CrossRef]

18. Jaśniok, T.; Jaśniok, M. Corrosion rate of reinforcement in concrete within a 3-year period of exposure at a monitored temperature and humidity. *Ochrona Przed Korozją* **2015**, *1*, 5–11. [CrossRef]

19. Jaśniok, T.; Jaśniok, M. A simple method of limiting polarization range during measurements of corrosion rate of reinforcement in concrete. *Ochrona Przed Korozją* **2016**, *59*, 210–213. [CrossRef]

20. Jaśniok, M. Investigation and modelling of the impact of reinforcement diameter in concrete on shapes of impedance spectra. *Procedia Eng.* **2013**, *57*, 456–465. [CrossRef]

21. Poursaee, A.; Hansson, C.M. Galvanostatic pulse technique with the current confinement guard ring: The laboratory and finite element analysis. *Corros. Sci.* **2008**, *50*, 2739–2746. [CrossRef]

22. Deus, J.M.; Freire, L.; Montemor, M.F.; Nóvoa, X.R. The corrosion potential of stainless steel rebars in concrete: Temperature effect. *Corros. Sci.* **2012**, *65*, 556–560. [CrossRef]

23. Blanco, G.; Bautista, A.; Takenouti, H. EIS study of passivation of austenitic and duplex stainless steels reinforcements in simulated pore solutions. *Cem. Concr. Compos.* **2006**, *28*, 212–219. [CrossRef]

24. Ceroni, F.; Cosenza, E.; Gaetano, M.; Pecce, M. Durability issues of FRP rebars in reinforced concrete members. *Cem. Concr. Compos.* **2006**, *28*, 857–868. [CrossRef]

25. Kim, H.Y.; Park, Y.H.; You, Y.J.; Moon, C.K. Short-term durability test for GFRP rods under various environmental conditions. *Compos. Struct.* **2008**, *83*, 37–47. [CrossRef]

26. Mufti, A.; Onofrei, M.; Benmokrane, B.; BaNthia, N.; Boulfiza, M.; Newhook, J.; Bakht, B.; Tadros, G.; Brett, P. Report on the studies of GFRP durability in concrete from field demonstration structures. In Proceedings of the 3rd International Conference on Composites in Construction—CCC, Lyon, France, 11–13 July 2005; pp. 11–13.

27. Wang, X.H.; Chen, B.; Gao, Y.; Wang, J.; Gao, L. Influence of external loading and loading type on corrosion behavior of RC beams with epoxy-coated reinforcements. *Constr. Build. Mater.* **2015**, *93*, 746–765. [CrossRef]

28. Tang, F.; Chen, G.; Brow, R.K. Chloride-induced corrosion mechanism and rate of enamel- and epoxy-coated deformed steel bars embedded in mortar. *Cem. Concr. Res.* **2016**, *82*, 58–73. [CrossRef]

29. Manning, D.G. Corrosion performance of epoxy-coated reinforcing steel: North American experience. *Constr. Build. Mater.* **1996**, *10*, 349–365. [CrossRef]

30. Figueira, R.M.; Pereira, E.V.; Silva, C.J.R.; Salta, M.M. Corrosion protection of hot dip galvanized steel in mortar. *Port. Electrochim. Acta* **2013**, *31*, 277–287. [CrossRef]

31. Kouril, M.; Pokorny, P.; Stoulil, J. Corrosion Mechanism and Bond-Strength Study on Galvanized Steel in Concrete Environment. *Corros. Sci. Technol.* **2017**, *16*, 69–75. [CrossRef]

32. Wojtynek, K.; Węgrzynkiewicz, S.; Kiełbus, A.; Sozańska, M.; Leclair, A. Zróżnicowanie grubości powłoki cynkowej na pierścieniu ochronnym do napowietrznych linii elektroenergetycznych. *Ochrona Przed Korozją* **2018**, *4*, 100–105. [CrossRef]

33. Nowicka-Nowak, M.; Zubielewicz, M.; Kania, H.; Liberski, P.; Sozańska, M. Influence of hot-dip galvanised coating morphology on the adhesion of organic coatings depending on the Zinc Bath Pb content and the postgalvanising cooling metod. *Int. J. Corros.* **2018**, *21*, 1–8. [CrossRef]

34. Langill, T.J.; Dugan, B. Zinc Materials for Use in Concrete. In *Galvanized Steel Reinforcement in Concrete*; Elsevier Science: Amsterdam, The Netherlands, 2004; pp. 87–109.

35. Swamy, R.N. Design for Durability with Galvanized Reinforcement. In *Galvanized Steel Reinforcement in Concrete*; Elsevier Science: Amsterdam, The Netherlands, 2004; pp. 31–69.

36. FIB Bulletin 49. *Corrosion Protection of Reinforcing Steels-Technical Report*; International Federation for Structural Concrete (FIB): Lausanne, Switzerland, 2009.

37. Andrade, C.; Alonso, C. Electrochemical Aspects of Galvanized Reinforcement Corrosion. In *Galvanized Steel Reinforcement in Concrete*; Elsevier: Amsterdam, The Netherlands, 2004; pp. 111–144. [CrossRef]

38. Tan, Z.Q.; Hansson, C.M. Effect of surface condition on the initial corrosion of galvanized reinforcing steel embedded in concrete. *Corros. Sci.* **2008**, *50*, 2512–2522. [CrossRef]

39. Pernicova, R.; Dobias, D.; Pokorny, P. Problems Connected with use of Hot-dip Galvanized Reinforcement in Concrete Elements. *Procedia Eng.* **2017**, *172*, 859–866. [CrossRef]

40. Roventi, G.; Bellezze, T.; Giuliani, G.; Conti, C. Corrosion resistance of galvanized steel reinforcements in carbonated concrete: Effect of wet–dry cycles in tap water and in chloride solution on the passivating layer. *Cem. Concr. Res.* **2014**, *65*, 76–84. [CrossRef]

41. Darwin, D.; Browning, J.; O'Reilly, M.; Xing, L.; Ji, J. Critical chloride corrosion threshold of galvanized reinforcing bars. *ACI Mater. J.* **2009**, *106*, 176–183.

42. EN 206+A1. *Concrete. Specification, Performance, Production and Conformity*; CEN-CENELEC: Brussels, Belgium, 2016.

43. Nürnberger, U.; Beul, W. Einfluß einer Feuerverzinkung und PVC-Beschichtung von Bewehrungsstählen und von Inhibitoren auf die Korrosion von Stahl in gerissenem Beton. *Werkstoffe Und Korrosion* **1991**, *42*, 537–546. [CrossRef]

44. Treadaway, K.W.J.; Brown, B.L.; Cox, R.N. ASTM STP 713. Durability of galvanized steel in concrete. In *Corrosion of Reinforcing Steel in Concrete*; ASTM: West Conshohocken, PA, USA, 1980; pp. 120–131.

45. Yeomans, S.R. Performance of black, galvanized and epoxy-coated reinforcing steel in chloride-contaminated con-crete. *Corrosion* **1994**, *50*, 72–81. [CrossRef]

46. Tonini, D.E. *ASTM STP 629. Deans S.W. Chloride Corrosion of Steel in Concrete*; ASTM: West Conshohocken, PA, USA, 1976.

47. EN 1766. *Products and Systems for the Protection and Repair of Concrete Structures. Test Methods. Reference Concretes for Testing*; CEN-CENELEC: Brussels, Belgium, 2017.

48. EN 1992-1-1. *Eurocode 2: Design of Concrete Structures-Part 1-1: General Rules and Rules for Buildings*; CEN-CENELEC: Brussels, Belgium, 2008.

49. ASTM A767/A767M. *Standard Specification for Zinc-Coated (Galvanized) Steel Bars for Concrete Reinforcement*; ASTM: West Conshohocken, PA, USA, 2015.

50. ASTM A123/A123M. *Standard Specifications for Zinc (Hot-Dipped Galvanized) Coatings on Iron and Steel Products*; ASTM: West Conshohocken, PA, USA, 2017.

51. Andrade, C.; Alonso, C. On-site measurements of corrosion rate of reinforcements. *Constr. Build. Mater.* **2001**, *15*, 141–145. [CrossRef]

52. Pawełek, M.; Węgrzynkiewicz, S.; Sozańska, M. Wpływ obróbki strumieniowo-ściernej na grubość i mikrostrukturę powłoki cynkowej utworzonej na powierzchni stali po wypalaniu gazowym. *Ochrona Przed Korozją* **2019**, *1*, 42–47. [CrossRef]

53. ASTM C876-91. *Standard Test Method for Half-Cell Potentials of Uncoated Reinforcing Steel in Concrete*; ASTM: West Conshohocken, PA, USA, 1999.

Article

Protective Geopolymer Coatings Containing Multi-Componential Precursors: Preparation and Basic Properties Characterization

Chenhui Jiang [1,2,3,4,*]**, Aiying Wang** [1,*]**, Xufan Bao** [2]**, Zefeng Chen** [4]**, Tongyuan Ni** [5] **and Zhangfu Wang** [4]

[1] Key Laboratory of Marine Materials and Related Technologies, Zhejiang Key Laboratory of Marine Materials and Protective Technologies, Ningbo Institute of Materials Technology and Engineering, Chinese Academy of Sciences, Ningbo 315201, China

[2] Hongrun Construction Group Co., Ltd., Shanghai 200235, China; xfbao_hr@126.com

[3] Department of Construction Engineering, Zhejiang College of Construction, Hangzhou 311231, China

[4] Zhejiang Huawei Building Materials Group, Co., Ltd., Hangzhou 310018, China; chenzf_11747@126.com (Z.C.); piter_wang1968@126.com (Z.W.)

[5] College of Civil Engineering, Zhejiang University of Technology, Hangzhou 310023, China; hznity@zjut.edu.cn

* Correspondence: chjiang2020@126.com (C.J.); aywang@nimte.ac.cn (A.W.)

Received: 2 July 2020; Accepted: 3 August 2020; Published: 5 August 2020

Abstract: This paper presents an experimental investigation on geopolymer coatings (GPC) in terms of surface protection of civil structures. The GPC mixtures were prepared with a quadruple precursor simultaneously containing fly ash (FA), ground granulated blast-furnace slag (GBFS), metakaolin (MK), and Portland cement (OPC). Setting time, compressive along with adhesive strength and permeability, were tested and interpreted from a perspective of potential applications. The preferred GPC with favorable setting time (not shorter than 120 min) and desirable compressive strength (not lower than 35 MPa) was selected from 85 mixture formulations. The results indicate that balancing strength and setting behavior is viable with the aid of the multi-componential precursor and the mixture design based on total molar ratios of key oxides or chemical elements. Adhesive strength of the optimized GPC mixtures was ranged from 1.5 to 3.4 MPa. The induced charge passed based on a rapid test of coated concrete specimens with the preferred GPC was 30% lower than that of the uncoated ones. Setting time of GPC was positively correlated with $\eta[Si/(Na+Al)]$. An abrupt increase of setting time occurred when the molar ratio was greater than 1.1. Compressive strength of GPC was positively affected by mass contents of ground granulated blast furnace slag, metakaolin and ordinary Portland cement, and was negatively affected by mass content of fly ash, respectively. Sustained seawater immersion impaired the strength of GPC to a negligible extent. Overall, GPC potentially serves a double purpose of satisfying the usage requirements and achieving a cleaner future.

Keywords: geopolymer coatings (GPC); setting time; compressive strength; adhesive strength; impermeability

---

1. Introduction

Geopolymers, also known as inorganic polymers or alkali-activated materials (AAMs), are a cluster of materials synthesized with alkali-activation and subsequent geopolymerization at elevated or ambient temperature [1–3]. Davidovits distinguished geopolymers from AAMs in that the both pertain to different systems, respectively, from the perspective of material science [4]. Due to industrial by-products and/or other inexpensive aluminosilicate sources such as fly ash (FA), granulated blast furnace slag (GBFS), steel slag, metakaolin (MK), waste glass, saw dust, mine tailing, rice husk ash, natural pozzolans, and water glass (WG) can be potentially utilized in production of geopolymers [5]; a large quantity of non-renewable resources stands a chance of conservation as well as massive solid waste is recycled and reused. Less greenhouse gas emission can be achieved by replacing traditional cement-based materials with geopolymer-based alternatives in building industry and civil engineering sector [6]. This means that geopolymer-based materials often lead to less energy consumption and environmental footprint.

Furthermore, geopolymer-based materials possess comparable or even superior mechanical properties and durability, especially excellent resistance to chemical corrosion and thermal properties [7–14]. Research findings revealed that geopolymers blended with various mineral additives show a significant improvement in mechanical properties and durability at various temperature conditions [7,8]. The results advocate that geopolymer mixtures with desired properties can be designed for ambient temperature curing condition with minerals additives, which may further promote them as an environmentally friendly construction material [7–9]. Geopolymer-based binders show inherently superior fire resistance as compared to Portland cement-based binders. However, it requires careful choice of precursor, use of aggregates, total alkali content in geopolymer, water content, and so forth [10]. The production of geopolymer concrete requires great care and correct material composition [10,11]. During the activation process in making the geopolymer, high alkalinity also requires safety risk and enhanced energy consumption and generation of greenhouse gases [11,14]. Geopolymer concretes using aggregates of different reactivity are reported to expand less than the corresponding Portland cement-based concretes [12,13]. The ability to utilize alkali-silica reaction vulnerable aggregate in the production of geopolymer concrete would increase economic and sustainability appeal [13,14].

Hence, as typical of cleaner and sustainable materials, geopolymers are expected to be powerful and cleaner products with promising prospects in the near future. They have been attracting more and more attention worldwide. Today the primary application of geopolymers in civil engineering is the development and production of building materials as a potential alternative to traditional materials with high energy consumption and pollution [15–18]. Geopolymers are expected to be green binding material and consume less energy. Production of geopolymer minimizes waste production and protect environment [15,16]. Geopolymers/alkali-activated binders have attracted considerable attention as promising construction and repair materials since their discovery because of their superior properties. Moreover, less pollution was caused by geopolymer/alkali-activated concretes than conventional cement concretes [17,18]. However, further research is needed to identify the sustainability and low-carbon nature of geopolymers and related products [1,3,5].

For the purpose of prolonging the lifespan of civil structures, sealing the exposed surface of concrete with coatings has been widely studied and utilized. Surface protection/treatment has become more significant in marine concrete structure, which is vulnerable to chemical deterioration and physical damage. Although organic coating materials (e.g., epoxy resins, silane and acrylic) have been commonly used, there exist disadvantages and drawbacks involving ease to crack/peel/degradation, inability to release vapor pressure [19], susceptible to fire [20], release of odors, lack of stability under UV radiation [21], difficulty to be removed after aging [22] (Pan et al., 2017), and low resistance against thermal shock [23]. In comparison, as previously reported by Franzoni [24] and Jiang [25], inorganic coatings possess desirable adhesive strength, good durability and favorable compatibility with concrete substrate. In addition, geopolymers show inherently superior fire and thermal shock resistance [26]. However, less attention and interest has been concentrated on this field, especially with regard to their penetration depth, bonding interface, interactions with substrate [22], and transport mechanism of deleterious ions in this type of coating layer.

As an innovative inorganic material with excellent performance, geopolymer coatings (GPC) can also be potentially used for the protective layer of concrete structures. Zhang et al. presented an experimental study of employing geopolymer-based material as a coating material for marine concrete protection [27]. Aguirre-Guerrero et al. prepared alkaline-activated FA/MK geopolymer mortar as an interfacial agent that prevented reinforcing bars embedded in reduced-scale concrete members from corrosion [28]. Wiyono reported that surface soundness of pozzolan concrete can be enhanced through the application of GPC layer [29]. Lv and co-authors synthesized a powder GPC by mixing MK and solid water glass along with additives and fillers [30]. They claimed that the dry-mixed coating material has been successfully applied in indoor engineering. Wang and Zhao issued a tentative study on silica fume-based GPC for fireproof plywood that was modified with decanoic-palmitic eutectic mixture [31]. In addition, from a point of view of sustainability, the carbon footprint of GPC is several times lower than that of traditional organic-based coatings [32]. The GPC prepared from multi-componential precursors at ambient temperature and its applications on concrete substrate, however, has seldom been studied, thus hindering a better understanding towards potential utilization of geopolymers.

The objective of this study, hence, is to design and prepare green GPC mixtures comprised with a quadruple precursor at ambient condition and to correlate basic engineering properties of GPC to its several key influencing factors. This paper is structured as follows: In the first step, a series of GPC were formulated and prepared based on the proposed methodology. Secondly, setting time and mechanical strength alongside rapid chloride permeability of the prepared and optimized GPC mixtures were experimentally investigated. The correlations between these properties and key mixture parameters of GPC were interpreted. Moreover, mechanical strength of GPC cured in standard condition and in artificial seawater was compared. In the end, the adhesive strength of GPC to a typical concrete substrate was evaluated with a portable apparatus. As a complement, the potential sustainability of this type of geopolymer-based applications was briefly reviewed.

2. Experimental Program

2.1. Materials

The constituent materials used can be classified into three parts in terms of their functions, i.e., aluminosilicate precursors (or binders), alkali activators and additives. The precursor includes FA, GBFS, MK and ordinary Portland cement (OPC) as per Chinese standards and technical specifications. All raw materials used in the study are supplied by Zhejiang Huawei Building Materials Group, Co., Ltd., in Hangzhou, China. Compared to one single precursor, the quadruple combinations were expected to complement each other's advantages with regard to chemical compositions, specific properties (e.g., particle size distribution and reactivity), economical cost, as well as sustainability. X-ray fluorescent (XRF) analysis (Olympus, Tokyo, Japan) was carried out to obtain oxide compositions of the precursors as shown in Table 1.

Table 1. Chemical compositions of precursors for geopolymer coatings.

Precursors	Oxide Compositions (Mass Fraction, %)									
	SiO_2	Al_2O_3	Na_2O	CaO	Fe_2O_3	MgO	TiO_2	K2O	H_2O	LOI
MK	45.12	42.40	0.15	9.11	0.76	0.09	1.37	0.19	0.12	0.35
FA	37.00	31.88	0.66	9.11	8.35	1.16	1.81	1.29	0.13	0.46
GBFS-A [1]	26.08	13.51	0.26	45.66	0.45	8.53	0.67	0.40	0.21	0.22
GBFS-B [1]	25.36	12.49	0.29	47.38	0.45	6.31	0.89	0.35	0.31	0.43
GBFS-C [1]	24.34	13.67	0.29	44.97	0.41	8.21	0.80	0.41	0.16	0.13
GBFS-D [1]	25.79	13.11	0.26	46.28	0.77	7.00	0.95	0.42	0.11	0.25
OPC	17.27	6.63	0.25	58.25	6.38	2.69	0.51	0.74	0.16	0.42

[1] GBFS-A, B, C, and D with roughly the same chemical compositions come from different iron works.

In order to make the GPC protective layer as compact as possible, the overall particle size distribution (PSD, Lab Synergy, New York, NY, USA) of the multi-componential precursors was optimized with the aid of EMMA (Version 1, Elkem, Olso, Norway) [33], a software package of particle packing calculation. Figures 1 and 2 respectively show scanning electron micrographs (SEM, Nanoscience, Phoenix, AZ, USA) and the laser-based PSD of the precursors. The FA mainly consists of spherical particles of glass, while the GBFS, MK and OPC are all composed of irregular and angular particles that are similar with previously published results [34–36]. The MK possesses the finest particles in all the precursors used, as clearly shown in Figures 1 and 2. It was expected to effectively fill spaces between coarser particles and densify the overall quadruple binder system. The GBFS and OPC with higher reactivity are beneficial to geopolymerization and development of mechanical properties of GPC at ambient temperature.

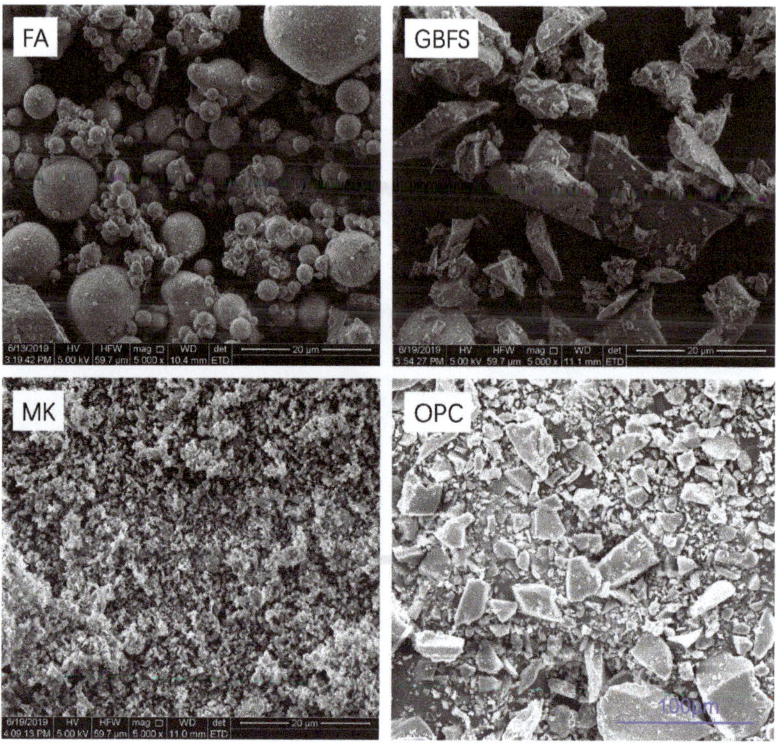

Figure 1. Scanning electron micrographs (SEM) of the precursors used.

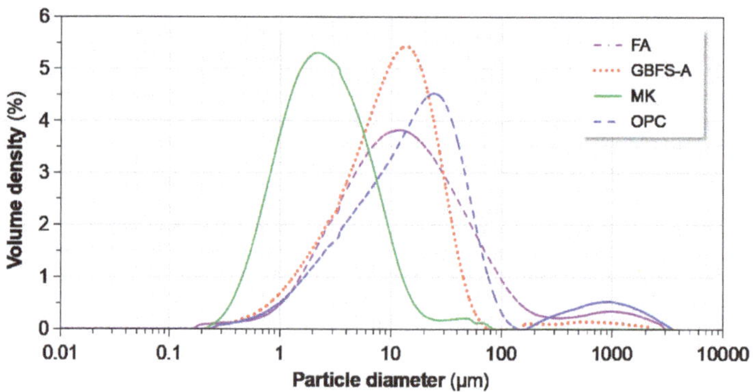

Figure 2. Particle size distributions (PSD) of the precursors employed in this study.

It is well-known that the geopolymerization process of geopolymer-based materials at ambient temperature is relatively slow and, therefore, a desirable strength is often difficult to obtain at early ages. Elevated temperature can accelerate the formation of geopolymer gel, and thus, enhance the strength. However, heat curing is often difficult to achieve in on-site applications. From the engineering and economical perspectives, therefore, it is essential to develop the GPC mixture with favorable mechanical strength under ambient condition. Based on the hypothesis that additional calcium silicate hydrate (C-S-H) gel coexists with geopolymer products, calcium-rich materials such as GBFS and OPC were introduced to shorten setting time and to improve mechanical properties of GPC. Additionally, heat release from hydration of the calcium-rich materials can accelerate the geopolymerization.

The alkaline activator was a blended aqueous solution of commercially available sodium hydroxide (NaOH pellets with 99.8% purity) and water glass (WG, with 27.5% SiO_2, 8.7% Na_2O and 63.8% H_2O by mass) synthesized from industrial by-products. A polycarboxylate-based superplasticizer (SP, with 26.2% solid mass content and 25.8% water-reducing ratio) and polypropylene fiber (PPF, with length of 6–10 mm and aspect ratio of 400–600) were used as functional additives of GPC. The former is favorable to improve workability of fresh GPC mixture while the latter is beneficial to toughness and cracking resistance of the hardened GPC layer. It is noteworthy that, once cracking occurs, a penetration pathway for corrosive agents will be emerged in the GPC layer and the protected substrate will be deteriorated.

2.2. Experimental Parameters

In this experimental investigation, primary engineering properties involving setting time, compressive and adhesive strength, along with permeability of a number of GPC mixtures were explored and interpreted. It is essential that GPC set neither too rapidly nor too slowly. In the first case, there might be insufficient time to operate the fresh mixture before it becomes too rigid. In the second case, too long a setting period tends to delay the work unduly, also it might postpone the actual use of the coating layer because of inadequate strength at the prescribed age. Consequently, initial and final setting time of GPC must be designated in a definite and reasonable range.

Compressive strength is the most frequently-used parameter for characterizing the engineering properties of hardened geopolymer-based materials [37,38]. Adhesive strength refers to the capability of a GPC layer to stick to a substrate. It is often measured by assessing the ultimate tensile stress to detach or unstick the hardened GPC layer perpendicular to the substrate. It is essential that GPC possesses robust impermeability to moisture and deleterious substances. As a typical scenario, the penetration and transport of chloride ions often resulted in deterioration of concrete structures in marine atmosphere. As a trial and error, we employed electric charge passed based on rapid chloride permeability test to characterize the protection effect of GPC.

It is noteworthy to state that all test data in this article are average values from two or three replicate samples on the basis of identical experimental configurations, respectively. Relative standard deviations (i.e., the ratio of standard deviation to average value) of all the presented data in figures and tables are in the range of 3–8%. The scatter of test data is totally lower and meets the requirements of experimental design. For these reasons, the error bars and statistical analyses of the experimental results were uniformly omitted.

2.3. Mixture Formulation

In order to achieve proper performance of end product of GPC, the synergy of precursor and alkaline activator requires to be fully taken into account. The procedures of mixture proportioning of GPC are schematically presented in Figure 3. For the sake of achieving the balance between setting time and mechanical strength, total molar ratios of SiO_2/Al_2O_3, SiO_2/Na_2O and H_2O/Na_2O as well as $Si/(Na+Al)$ in the GPC system were calculated and regulated. According to Davidovits [39,40] and Rangan [41], the geopolymer pastes with these molar ratios in the ranges of 3.3–4.5, 0.8–2.2 and 10–25 are suitable for usage of protective coatings, respectively. The additives such as SP, PPF and retarder were adopted to further improve the particular performance of the GPC mixtures. Due to the mass proportion of the precursor being predominant, the economic cost and sustainability of GPC depend on the quadruple compositions of precursors. Less MK and more FA were suggested.

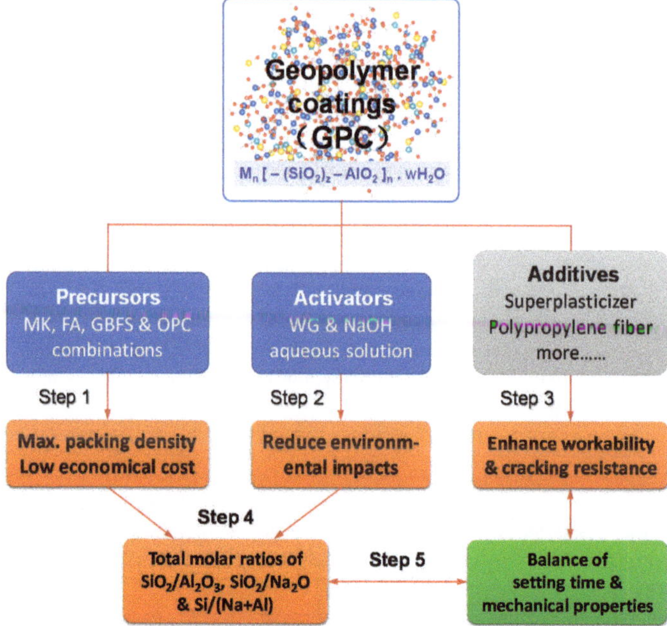

Figure 3. Mixture design procedure of GPC: a step by step methodology of mixture formulation involving particle packing, environmental impacts, engineering performance, as well as material chemistry.

Based on the principles of mixture design, a series of GPC mixtures (85 groups) with trial and error were examined. The mixtures were listed in Table 2. As stated previously, the GPC mixtures containing multi-componential precursors were highlighted. It should be noted that, in Section 3, the experimental results will be interpreted in terms of various factors that affected the properties rather than serial numbers of the GPC mixtures.

2.4. Mixture Preparation

The preparation of the GPC mixtures in the laboratory consists of the following four steps. First of all, prescribed quantities of WG and NaOH were carefully blended and dissolved in the mixing water to obtain the activator solution with the required concentration. The desired concentration (or molarity) of NaOH is suggested to be within the range of 8–16 mol/L. However, 8–10 mol/L is adequate for most cases [38,42]. Too high alkalinity is prone to result in the efflorescence [43] and safety hazard of operations. The activator solution was cooled down and stabilized for at least 24 h before further mixing with the precursors. Secondly, the prescribed MK, GBFS, FA, and OPC combinations were fully mixed together to ensure the uniformity using a mechanical mixer. In the third step, the formulated alkaline solution was fully mixed with the specific additives. At this point, the liquid and solid components of the GPC mixture were prepared. Finally, the solid component was added to the liquid one and then the mixing process was performed to obtain the fresh GPC mixtures. All these steps were performed with constant temperature and relative humidity (RH) of 25 °C ± 2 °C, 60% ± 5%, respectively.

Table 2. Mixture proportions of geopolymer coatings.

Mixture No.	Precursors (g)				Activators (g)			Additives (g)		W/S	Total Molar Ratios				
	MK	FA	GBFS	OPC	WG	NaOH	W	SP	PPF		SiO_2/Al_2O_3	SiO_2/Na_2O	Na_2O/Al_2O_3	$Si/(Na+Al)$	H_2O/Na_2O
01	250	0	0	250	355.1	53.6	84.5	7.50	0.25	0.450	3.513	3.417	1.028	0.866	13.86
02	250	250	0	0	355.1	53.6	84.5	7.50	0.25	0.450	3.195	4.136	0.772	0.901	13.55
03	250	250	0	0	355.1	53.6	132.5	7.50	0.25	0.520	3.195	4.136	0.772	0.901	15.66
04	250	125	0	125	355.1	53.6	132.5	7.50	0.25	0.520	3.330	3.781	0.881	0.885	15.84
05	125	250	0	125	355.1	53.6	132.5	7.50	0.25	0.520	3.881	3.672	1.057	0.943	15.64
06	250	0	250	0	355.1	53.6	132.5	7.50	0.25	0.520	3.349	3.665	0.914	0.875	15.81
07	250	0	250	0	355.1	53.6	132.5	7.50	0.25	0.520	3.349	3.665	0.914	0.875	15.81
08	250	0	250	0	355.1	53.6	132.5	7.50	0.25	0.520	3.388	3.685	0.919	0.883	16.00
09	250	0	250	0	355.1	53.6	132.5	7.50	0.25	0.520	3.286	3.650	0.900	0.865	16.00
10	250	0	250	0	355.1	53.6	132.5	7.50	0.25	0.520	3.363	3.703	0.908	0.881	16.01
11	0	500	0	0	355.1	53.6	132.5	7.50	0.25	0.520	3.011	3.699	0.814	0.830	15.56
12	0	400	100	0	355.1	53.6	132.5	7.50	0.25	0.520	3.271	3.574	0.915	0.854	15.63
13	0	300	200	0	355.1	53.6	132.5	7.50	0.25	0.520	3.610	3.447	1.047	0.882	15.71
14	0	200	300	0	355.1	53.6	132.5	7.50	0.25	0.520	4.068	3.320	1.225	0.914	15.79
15	0	100	400	0	355.1	53.6	132.5	7.50	0.25	0.520	4.722	3.191	1.480	0.952	15.88
16	50	200	200	50	355.1	53.6	132.5	7.50	0.25	0.520	3.755	3.390	1.108	0.891	15.81
17	0	200	200	100	355.1	53.6	132.5	7.50	0.25	0.520	4.202	3.203	1.312	0.909	15.80
18	0	300	100	100	355.1	53.6	132.5	7.50	0.25	0.520	3.695	3.331	1.109	0.876	15.72
19	0	100	300	100	355.1	53.6	132.5	7.50	0.25	0.520	4.944	3.073	1.609	0.948	15.88
20	0	400	0	100	355.1	53.6	132.5	7.50	0.25	0.520	3.327	3.458	0.962	0.848	15.64
21	0	0	400	100	355.1	53.6	132.5	7.50	0.25	0.520	6.135	2.942	2.085	0.994	15.96
22	125	125	125	125	355.1	53.6	132.5	7.50	0.25	0.520	3.661	3.400	1.077	0.881	15.89
23	50	200	150	100	355.1	53.6	132.5	7.50	0.25	0.520	3.804	3.332	1.142	0.888	15.81
24	100	200	150	50	355.1	53.6	132.5	7.50	0.25	0.520	3.462	3.519	0.984	0.873	15.82
25	50	150	150	150	355.1	53.6	132.5	7.50	0.25	0.520	4.118	3.209	1.283	0.902	15.85
26	75	250	150	25	525.1	45.5	76.1	7.50	0.25	0.559	4.000	3.894	1.027	0.987	17.00
27	75	200	200	25	466.2	44.9	100.9	7.50	0.25	0.558	4.000	3.897	1.026	0.987	17.70
28	75	150	250	25	407.4	44.3	146.9	7.50	0.25	0.588	4.000	3.899	1.026	0.987	19.54
29	75	100	300	25	348.5	43.8	150.4	7.50	0.25	0.557	4.000	3.902	1.025	0.988	19.47
30	75	0	400	25	230.8	42.6	200.0	7.50	0.25	0.555	4.000	3.911	1.023	0.989	21.99
31	75	200	175	50	459.5	44.3	81.8	7.50	0.25	0.528	4.000	3.895	1.027	0.987	16.90
32	75	175	200	50	430.1	44.0	94.7	7.50	0.25	0.528	4.000	3.897	1.026	0.987	17.28
33	75	150	225	50	400.6	43.8	107.4	7.50	0.25	0.527	4.000	3.898	1.026	0.987	17.70

Table 2. Cont.

Mixture No.	Precursors (g)				Activators (g)			Additives (g)		W/S	Total Molar Ratios				
	MK	FA	GBFS	OPC	WG	NaOH	W	SP	PPF		SiO_2/Al_2O_3	SiO_2/Na_2O	Na_2O/Al_2O_3	$Si/(Na+Al)$	H_2O/Na_2O
34	75	125	250	50	371.2	43.5	120.2	7.50	0.25	0.527	4.000	3.899	1.026	0.987	18.14
35	75	100	275	50	341.8	43.2	132.9	7.50	0.25	0.526	4.000	3.901	1.025	0.987	18.63
36	75	100	275	50	341.8	24.3	132.9	7.50	0.25	0.542	4.000	5.034	0.795	1.114	24.04
37	75	275	125	25	466.2	44.9	94.9	7.50	0.25	0.550	3.701	3.991	0.927	0.960	17.37
38	100	250	125	25	466.2	44.9	94.9	7.50	0.25	0.550	3.657	4.024	0.909	0.958	17.40
39	125	225	125	25	466.2	44.9	94.9	7.50	0.25	0.550	3.614	4.058	0.891	0.956	17.43
40	125	225	125	25	584.8	20.6	29.5	7.50	0.25	0.550	4.000	5.046	0.793	1.116	20.09
41	125	225	125	25	520.0	32.0	64.0	7.50	0.25	0.550	3.789	4.567	0.830	1.035	18.87
42	125	225	125	25	540.0	28.0	53.0	7.50	0.25	0.550	3.854	4.734	0.814	1.062	19.32
43	125	225	125	25	570.0	26.0	39.0	7.50	0.25	0.550	3.952	4.782	0.826	1.082	19.28
44	125	225	125	25	575.0	24.0	36.0	7.50	0.25	0.550	3.968	4.877	0.814	1.094	19.59
45	125	225	125	25	461.4	20.1	83.4	7.50	0.25	0.550	3.598	5.411	0.665	1.081	22.48
46	150	200	125	25	466.2	44.9	94.9	7.50	0.25	0.550	3.572	4.091	0.873	0.954	17.46
47	175	175	125	25	466.2	44.9	94.9	7.50	0.25	0.550	3.532	4.125	0.856	0.951	17.49
48	125	240	125	10	606.5	20.4	19.9	7.50	0.25	0.550	4.000	5.045	0.793	1.116	19.78
49	75	250	150	25	0.0	100.0	300.0	7.50	0.25	0.500	2.163	2.201	0.983	0.546	12.97
50	75	250	150	25	0.0	100.0	330.0	7.50	0.25	0.550	2.163	2.201	0.983	0.546	14.26
51	125	225	125	25	584.8	23.4	31	7.50	0.25	0.550	4.000	4.890	0.818	1.100	19.54
52	125	225	125	25	554.1	21.8	43.6	7.50	0.25	0.550	3.900	5.045	0.773	1.100	20.32
53	125	225	125	25	523.4	20.1	56.2	7.50	0.25	0.550	3.800	5.226	0.727	1.100	21.23
54	125	225	125	25	492.7	18.5	68.6	7.50	0.25	0.550	3.700	5.424	0.682	1.100	22.22
55	125	225	125	25	462	16.8	81.3	7.50	0.25	0.550	3.600	5.658	0.636	1.100	23.39
56	125	225	125	25	431.3	15.2	93.9	7.50	0.25	0.550	3.500	5.920	0.591	1.100	24.70
57	50	200	200	50	414.9	24.6	106.3	7.50	0.25	0.550	4.000	4.890	0.818	1.100	22.30
58	75	100	275	50	0	0	275	7.50	0.25	0.550	2.468	97.381	0.025	1.204	22.33
59	75	100	275	50	400	40	121	7.50	0.25	0.550	4.261	3.994	1.067	1.031	19.19
60	75	100	275	50	400	35	118.5	7.50	0.25	0.550	4.261	4.237	1.006	1.062	20.23
61	75	100	275	50	400	30	116	7.50	0.25	0.550	4.261	4.512	0.944	1.096	21.40
62	75	100	275	50	400	25	113	7.50	0.25	0.550	4.261	4.825	0.883	1.131	22.69
63	75	100	275	50	400	20	110	7.50	0.25	0.550	4.261	5.184	0.822	1.169	24.19
64	75	100	275	50	450	25	91	7.50	0.25	0.550	4.485	4.711	0.952	1.149	21.62
65	75	100	275	50	450	25	91	7.50	0.25	0.550	4.485	4.711	0.952	1.149	21.62

Table 2. *Cont.*

Mixture No.	Precursors (g)				Activators (g)			Additives (g)		W/S	Total Molar Ratios				
	MK	FA	GBFS	OPC	WG	NaOH	W	SP	PPF		SiO_2/Al_2O_3	SiO_2/Na_2O	Na_2O/Al_2O_3	Si/(Na+Al)	H_2O/Na_2O
66	75	100	275	50	500	25	69	7.50	0.25	0.550	4.709	4.612	1.021	1.165	20.68
67	75	100	275	50	550	25	47	7.50	0.25	0.550	4.933	4.526	1.090	1.180	19.87
68	75	100	275	50	600	25	25	7.50	0.25	0.550	5.157	4.450	1.159	1.194	19.15
69	75	100	275	50	600	25	25	7.50	0.25	0.550	5.157	4.450	1.159	1.194	19.15
70	75	100	275	50	400	42.7	122.8	7.50	0.25	0.550	4.261	3.874	1.100	1.015	18.71
71	75	100	275	50	400	45.1	124.1	7.50	0.25	0.550	4.261	3.773	1.129	1.001	18.28
72	75	100	275	50	400	46.4	124.8	7.50	0.25	0.550	4.261	3.721	1.145	0.993	18.06
73	75	100	275	50	400	17	108	7.50	0.25	0.550	4.261	5.427	0.785	1.193	25.18
74	75	100	275	50	400	15.5	107.8	7.50	0.25	0.550	4.261	5.557	0.767	1.206	25.77
75	75	100	275	50	400	12.1	106	7.50	0.25	0.550	4.261	5.876	0.725	1.235	27.11
76	75	100	275	50	419.9	24.4	104	7.50	0.25	0.550	4.350	4.816	0.903	1.143	22.41
77	75	100	275	50	404.3	24.2	110.8	7.50	0.25	0.550	4.280	4.868	0.879	1.139	22.83
78	75	100	275	50	400	0	99	7.50	0.25	0.550	4.261	7.384	0.577	1.351	33.41
79	75	100	275	50	450	0	77.3	7.50	0.25	0.550	4.485	6.943	0.646	1.362	30.70
80	75	100	275	50	500	0	55.4	7.50	0.25	0.550	4.709	6.586	0.715	1.373	28.50
81	75	100	275	50	550	0	33.4	7.50	0.25	0.550	4.933	6.293	0.784	1.383	26.69
82	75	100	275	50	600	0	11.5	7.50	0.25	0.550	5.157	6.047	0.853	1.392	25.17
83	75	100	275	50	400	100	154.5	7.50	0.25	0.550	4.261	2.365	1.802	0.760	12.38
84	75	100	275	50	400	5	102	7.50	0.25	0.550	4.261	6.676	0.638	1.300	30.46
85	75	100	275	50	500	85	102	7.50	0.25	0.550	4.709	2.682	1.756	0.854	13.05

2.5. Experimental Procedures

Initial and final set of GPC mixtures was determined with Vicat needle method, which is in accordance with ASTM C191 [44] with necessary modifications [45]. The method requires casting the fresh mixture into a truncated ring with a height of 40 mm. The measurement started after 10–15 min of curing the GPC samples in an environmental chamber (25 °C ± 2 °C and 95% ± 5% RH) and continued with an interval of 5–15 min up to the final set.

Compressive strength of GPC was performed on 20 mm cubes at the ages of 3, 7 (or 10) and 28 days, respectively. The compacted specimens (six replicates for each test) with the steel molds were initially stored in the above-mentioned chamber (standard condition) until 24 h and then stripped, and placed again in standard condition up to the measured ages. The tests were performed on a mechanical testing machine with a capacity of 300 kN. To examine the latent influence of curing condition on compressive strength, a comparison group for each GPC mixture was continuously immersed in saline water until the test ages. The saline water was prepared with crude sea salt of a 7% concentration that is an analogue of marine environment.

Adhesive strength (more accurately, pull-off adhesion strength) of GPC was determined using a portable adhesion tester (HCTC-10, provided by Beijing HICHANCE Technology Co., Ltd. in China). This test method covers procedures for evaluating the adhesion of a coating on concrete. The test determines the greatest perpendicular force in tension that a surface area can bear before a plug of material is detached. Failure will occur along the weakest plane within the system comprised of the loading fixture, glue, coating system, and substrate, and will be exposed by the fracture surface. The fresh GPC was painted and coated on the surface of concrete substrate (with an area of 100 mm × 400 mm) that was previously polished and handled to a saturated and surface dry state. The thickness of GPC layer was 3 mm. As shown in Figure 4, after being exposed to standard conditions up to 5 and 26 days, five standard steel flanges with Φ50 mm smooth surface were glued on the GPC layer with epoxy resin (Figure 4a). After the resin was fully cured (~48 h), the flanges were detached using a portable tester and adhesive strength was displayed on the LCD screen (Figure 4b). The failure interface in Figure 4c indicated that adhesive strength of GPC is comparably higher than that of the mortar part in concrete. In addition, the concrete substrate used was previously fabricated with the following mixture proportion (kg/m^3): 255 kg OPC, 206 kg FA, 846 kg river sand, 928 kg crushed stone, 160 kg water, and 9.5 kg SP. The concrete substrate was cast and cured in standard condition until the age of 28 days. The characteristic value of compressive strength is 40 MPa.

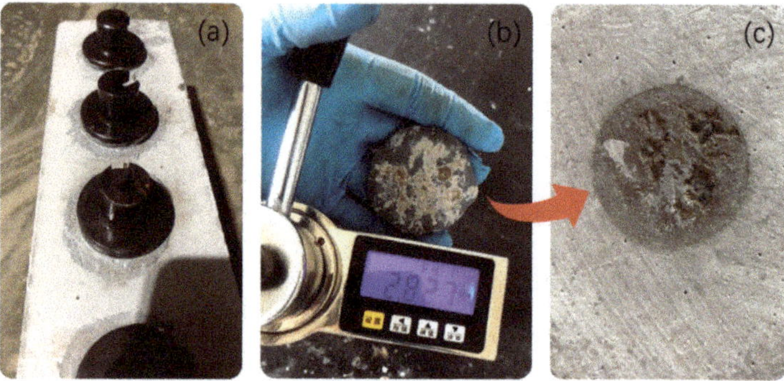

Figure 4. Adhesive strength test of GPC: (**a**) standard steel flanges glued on the surface of GPC layer that was previously coated on concrete substrate; (**b**) pulling out steel flanges from glued GPC surface with a portable adhesion tester, the LCD screen displays adhesive strength (MPa) of GPC; (**c**) failure interface between GPC layer and concrete substrate.

As the initial attempt, chloride permeability of GPC characterized through charge passed (Coulombs) was tested with a rapid method as per ASTM C1202 standard [46]. One cross section of the protected cylinder concrete specimen ($\Phi 100 \times 50$ mm) was previously polished and coated with the GPC mixture to be examined. At the age of 28 days, the charge passed of the coated concrete specimen was determined with a specialized apparatus [46] in parallel with the control one without GPC.

3. Results and Discussion

3.1. Setting Time

The GPC mixtures solely prepared with GBFS precursor often give high early-age strength but very rapid set within 30 min. Commercial retarders for OPC have been proven ineffective in geopolymer systems [47]. Lv and Cui [30] reported that an addition of polycondensed aluminum phosphate can prolong the solidification of an inorganic geopolymer-based coatings. However, the authors were unable to achieve the observation in this study. It seems as if there is a dilemma between strength development and setting time of GPC. Replacing GBFS with other aluminosilicates and formulating multi-component precursors can prolong setting time to great extent. In this section, we will present and interpret a series of factors affecting the GPC's setting time.

3.1.1. Effect of Activator Composition

The influence of the alkaline activator on setting time of GPC mixtures can be separately explored from its triple compositional makeup, viz., the mass of water glass, NaOH and additional water. The experimental results are clearly plotted in Figure 5 respectively. In these scatter diagrams, the test data is based on the identical precursor combinations (MK = 125 g, FA = 225 g, GBFS = 125 g and OPC = 25 g) and the constant water-solid ratio (W/S = 0.55).

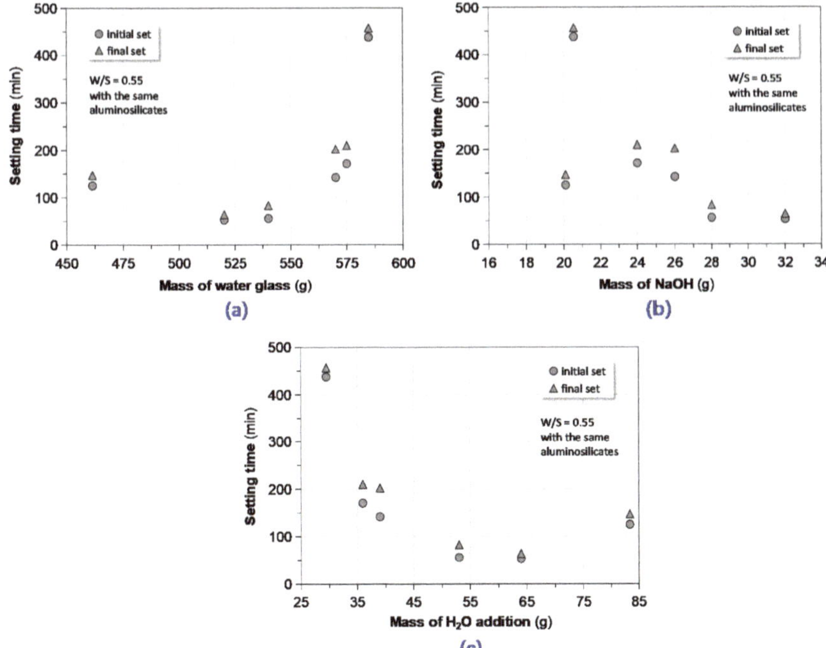

Figure 5. Correlating setting time to activator solutions of GPC: (**a**) setting time vs. amount of water glass; (**b**) setting time vs. amount of NaOH; (**c**) setting time vs. amount of additional mixing water.

In general, the relationships between the setting time and the activator composition were not quite monotonous. The intervals between the initial and final setting time of different mixtures were not long enough as those of the Portland cement paste did. However, for a first approximation, more water glass led to the extension of setting time while an increase of the quantity of NaOH or the mixing water resulted in a decrease of setting time, respectively. Keep in mind that the alkalinity and the pH value of NaOH is much higher than water glass; their contributions on the activation and geopolymerization are different; and in turn the influence of setting time is somewhat distinguished. With regard to the mixing water, its impact manner and degree cannot seem to be explained with the dilution effect that is applicable to cement-based materials, even though the polycondensation of geopolymer produces water molecules [15,17,23,29].

As far as the effect of the precursors on setting time was concerned, it was observed that the style or mode of action of each component in the quadruple precursor failed to come to any agreement. Due to the fact that GPC is a complex mixture system, we should interpret physical mechanisms of various factors on its property from a more comprehensive viewpoint. Maybe, replacing the single components as impact factors with the total quantities/moles of specific oxides or chemical elements in the entire GPC system is more reasonable [23,26].

3.1.2. Effect of Key Molar Ratios

In this paper, the total molar ratios of several key oxides (i.e., SiO_2, Na_2O, Al_2O_3, and H_2O) or chemical elements (i.e., Si, Na and Al) are denoted as $\eta(SiO_2/Na_2O)$, $\eta(SiO_2/Al_2O_3)$, $\eta(Na_2O/Al_2O_3)$, $\eta(H_2O/Na_2O)$, and $\eta[Si/(Na+Al)]$. In Figures 6 and 7, the authors tried to correlate setting time of the GPC mixtures to the total molar ratios of these oxides or elements. $\eta(SiO_2/Na_2O)$ and $\eta(SiO_2/Al_2O_3)$ separately represented the alkali and silica contents in the GPC system.

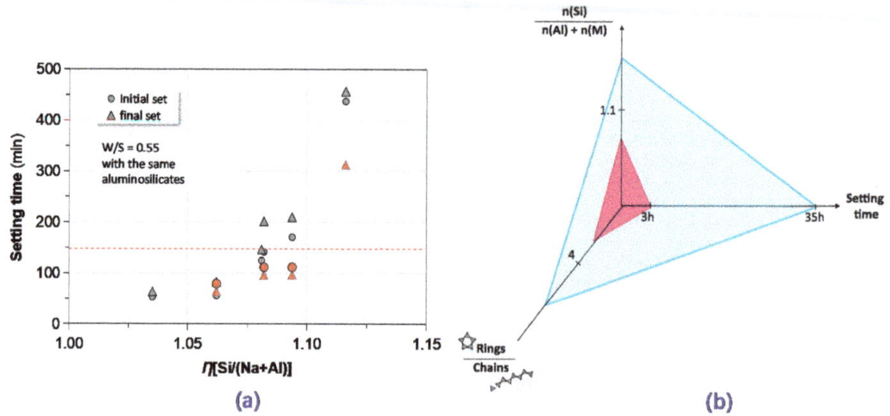

Figure 6. Correlating $\eta[Si/(Na+Al)]$ to setting time of GPC: (**a**) setting time vs. $\eta[Si/(Na+Al)]$ in this study alongside the reported data; (**b**) setting time vs. $\eta[Si/(Na+Al)]$ reported in previous literature [48].

According to the interrelation among setting time, $\eta[Si/(Na+Al)]$ and alkaline silicate's molecule structure (ring or chain) of MK-based geopolymer developed by Arnoult et al. [48], as shown in Figure 6b, with a reference value of $\eta[Si/(Na+Al)] = 1.1$, were selected to explore a potential relationship between setting time and $\eta[Si/(Na+Al)]$ of the GPC mixtures. The results are depicted in Figure 6a along with the data points collected from preceding studies [47,48]. It was obvious that the setting time was positively correlated with $\eta[Si/(Na+Al)]$. Once the value of $\eta[Si/(Na+Al)]$ exceeded 1.1, an abrupt increase of setting time occurred. The default value of $\eta[Si/(Na+Al)]$, which corresponds to favorable setting time (100–200 min), was seemingly suitable for the current coating applications.

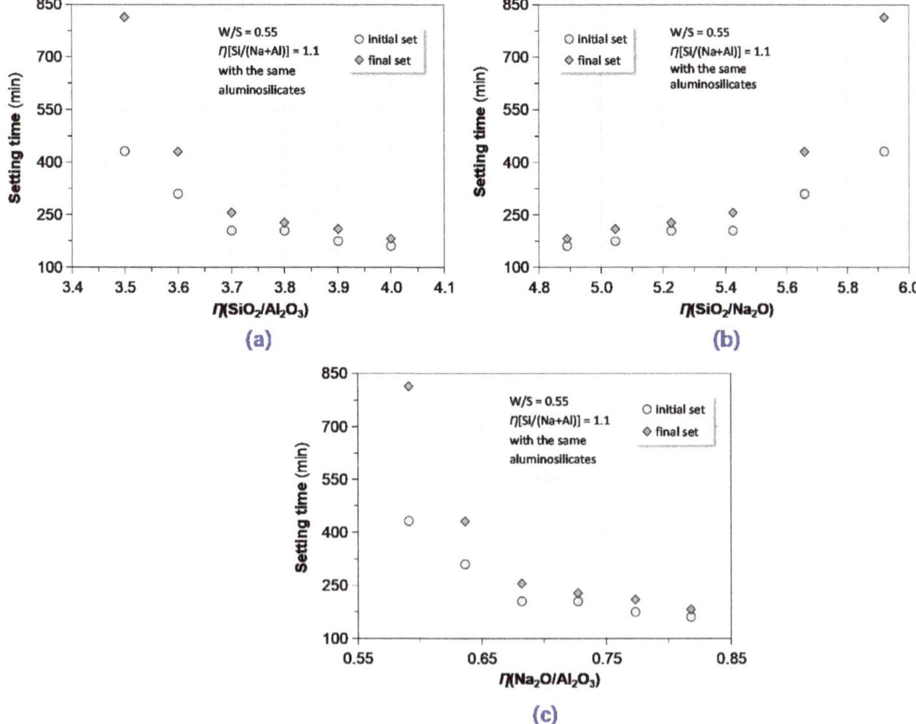

Figure 7. Correlating setting time to key molar ratios of GPC: (**a**) setting time vs. $\eta(SiO_2/Al_2O_3)$; (**b**) setting time vs. $\eta(SiO_2/Na_2O)$ and (**c**) setting time vs. $\eta(Na_2O/Al_2O_3)$.

Keep a constant value of $\eta[Si/(Na+Al)] = 1.1$ and the identical precursors; the relationships between setting time and the above-mentioned total molar ratios of key oxides are presented in Figure 7. Apparently, compared to the preceding correlations concerning setting time and single components, the setting time-total molar ratio relationships were definite and straightforward. It is also worth noting that the interval between initial and final setting time gradually increased with a decrease of $\eta(SiO_2/Al_2O_3)$ or $\eta(Na_2O/Al_2O_3)$ as well as an increase of $\eta(SiO_2/Na_2O)$, respectively.

Ozer and Soyer-Uzun [49] characterized the effect of $\eta(Si/Al)$ on reaction products of alkali-activated MK with X-ray diffraction (XRD) and SEM technology. They found that the geopolymer product transformed from crystalline phase (zeolite-A) to amorphous microstructures (N-A-S-H gel) with the increase of $\eta(Si/Al)$. The observation was recently further supported by Chen et al. [50]. The extension of setting time with the reduction of $\eta(SiO_2/Al_2O_3)$ was probably ascribed to this phase transformation, but the latent mechanism is still unclear. In the present usage of geopolymer, in order to obtain proper setting behavior, the appropriate $\eta(SiO_2/Al_2O_3)$ should be selected from the range of 3.7–4.0.

3.1.3. Effect of Preparation Method

It was generally suggested that the blended activator solution (water glass + alkali metal hydroxides + H_2O) should be prepared at least 24 h in advance followed by mixed with the precursors. However, the reason why this emphasizes the procedure was not definitely stated. We took two sets of experiments randomly on two GPC mixtures to examine the effect of preparation of alkali-activator on basic properties such as setting time and compressive strength. In the control test, the activator was made as per the common procedures. Meanwhile, the mixed activator solution was immediately blended and stirred with the precursors in the parallel test. The results presented in Figure 8 indicated that the activator after stabilizing 24 h led to a more prolonged setting time than its immediately mixed counterpart did. In other word, after the storage of a period of time, the reactivity of the activator reduced to some extent. This assertion would also be explained from the results of early-age mechanical strength of the both GPC mixtures made from the activators with different initial states: the GPC mixtures made from the immediately mixed activator obtained higher compressive strength at the ages of 3 and 7 days. The effects of the preparation method on the strength and setting time were also dependent upon the precursor combinations along with the specific age. The effect on the strength was more obvious at an earlier age [35,44].

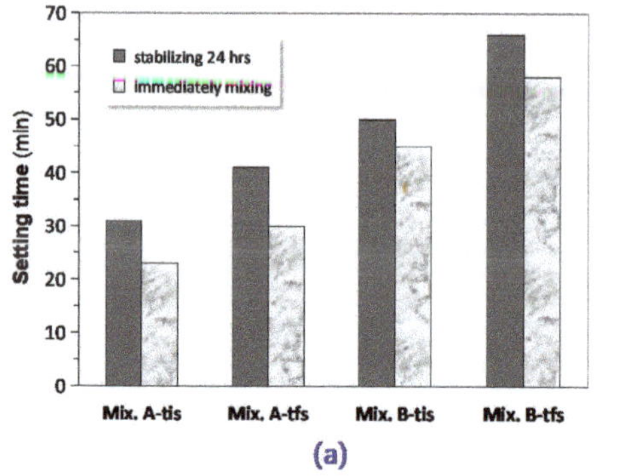
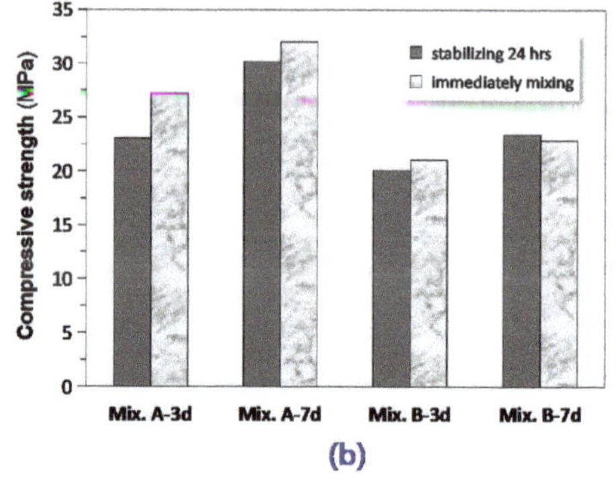

Figure 8. Influence of preparation procedure on basic properties of GPC: (**a**) setting time (tis = initial setting and tfs = final setting); (**b**) compressive strength at 3 and 7 days. In the two graphs, Mix. A and B stand for the GPC mixtures randomly extracted from the investigated mixtures.

3.2. Compressive Strength

Mechanical strength of geopolymer presented in the literature varied widely because of the specific attributes of constitute materials, methods of preparation and exposed conditions [51]. However, there is very limited research that has been performed on comprehensive evaluation of the influencing parameters. Thermal curing is often difficult to achieve and inconvenient for in-situ applications; therefore, it is desirable to develop a GPC that can come into effect at ambient temperature. Furthermore, what we found was that, in comparison with the influence of alkaline activator, the dependence between compressive strength and precursor composition of the GPC mixtures was much stronger and more robust.

3.2.1. Effect of Precursor Composition

As the binder of the geopolymer, the precursors and their make-up are of vital importance to strength of GPC. Since GBFS, FA and MK have been widely utilized to synthesize geopolymers, we proposed a blended precursor system involving these typical aluminosilicates along with an addition of OPC, supposing that their advantages are complementary. The respective proportions of these aluminosilicates in the quadruple precursor system were intended to optimize on the basis of compressive strength, more accurately the balance of strength and setting time of the GPC mixtures [21,35,44].

As shown in Figure 9, compressive strength of the GPC at various ages (3, 7, 10 and 28 days) was positively and negatively affected by the mass content of GBFS and FA in the quadruple precursor, respectively, regardless of activator's composition, W/S and curing condition. Keeping identical mass of (MK + FA) and GBFS, with an increase of mass content of MK, the GPC's compressive strength increased. The influences of these binder components on compressive strength strongly pertain to their chemical reactivity. The mass content ranges of the precursor components were normalized on the one hand by the particle packing and were limited on the other hand by the attributes. For instance, more than 35% MK will result in an increase of mixing water while more than 50% GBFS will lead to very rapid set. Therefore, when one formulates the GPC mixture, it is critical to balance the strength and setting properties.

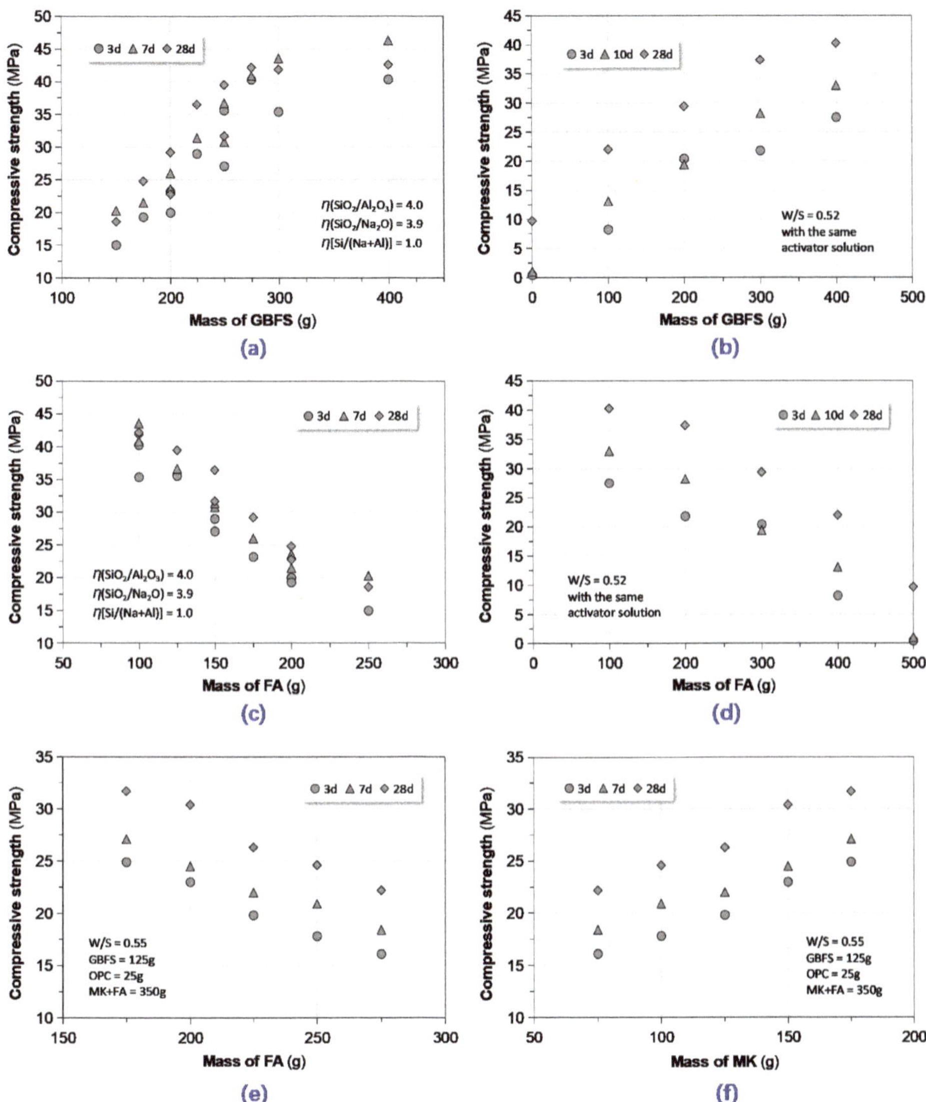

Figure 9. Correlating compressive strength to quadruple precursor composition of GPC: (**a,b**) compressive strength vs. mass of GBFS; (**c–e**) compressive strength vs. mass of FA; and (**f**) compressive strength vs. mass of MK.

3.2.2. Effect of Key Molar Ratios

According to our observations, the effects of key molar ratios on compressive strength were less evident than on setting time did. With regard to the influences of $\eta(SiO_2/Na_2O)$ and $\eta(SiO_2/Al_2O_3)$, several investigations have been carried out. Cheng et al. reported that the MK-waste catalyst-based geopolymer samples with higher $\eta(SiO_2/Na_2O)$ appeared to possess more compact structures, higher strength and lower porosity [52]. Gao et al. highlighted that the nano-SiO_2 and MK-based geopolymer with the $\eta(SiO_2/Na_2O)$ value of 1.5 exhibited higher strength and less porosity [53]. Wang et al. presented a comprehensive study on molecular structure and mechanical properties of geopolymers with different $\eta(SiO_2/Al_2O_3)$ by means of experiments and simulation. They indicated that, with the increases of $\eta(SiO_2/Al_2O_3)$, both the stability of molecular structure and mechanical properties of resultant geopolymer decreased [31]. Various zeolites were main products in geopolymers with a relatively low $\eta(SiO_2/Al_2O_3) = 2.0$–2.5. The formation of these zeolitic structures resulted in low-strength MK-based geopolymers [54]. These findings basically coincided with the relevant results in this work, so there is no need to repeat.

In the case of the effects of $\eta(H_2O/Na_2O)$, $\eta[Si/(Na+Al)]$ and W/S, the corresponding experimental results are plotted in Figure 10. With the identical $\eta(SiO_2/Na_2O)$, $\eta(SiO_2/Al_2O_3)$ and $\eta[Si/(Na+Al)]$, the correlation between compressive strength and W/S was elusive. The observation was basically different from the findings reported by Juengsuwattananon et al. who indicated that the geopolymer's strength decreased with an increase of W/S [54]. They ascribed the phenomenon to dilution effect by adding more water to the system. More mixing water led to a reduction in pH value of the system and adversely affected the reaction rate. In nature, W/S or liquid-solid ratio are the practical forms of $\eta(H_2O/Na_2O)$. However, the influences of $\eta(H_2O/Na_2O)$ on strength varied with experimental configurations. When $\eta(SiO_2/Na_2O)$, $\eta(SiO_2/Al_2O_3)$ and $\eta[Si/(Na+Al)]$ were held constant, higher $\eta(H_2O/Na_2O)$ led to greater strength. Otherwise, when the same precursors were used, the strength slightly decreased with the increase of $\eta(H_2O/Na_2O)$. The effect of $\eta[Si/(Na+Al)]$ was also inapparent.

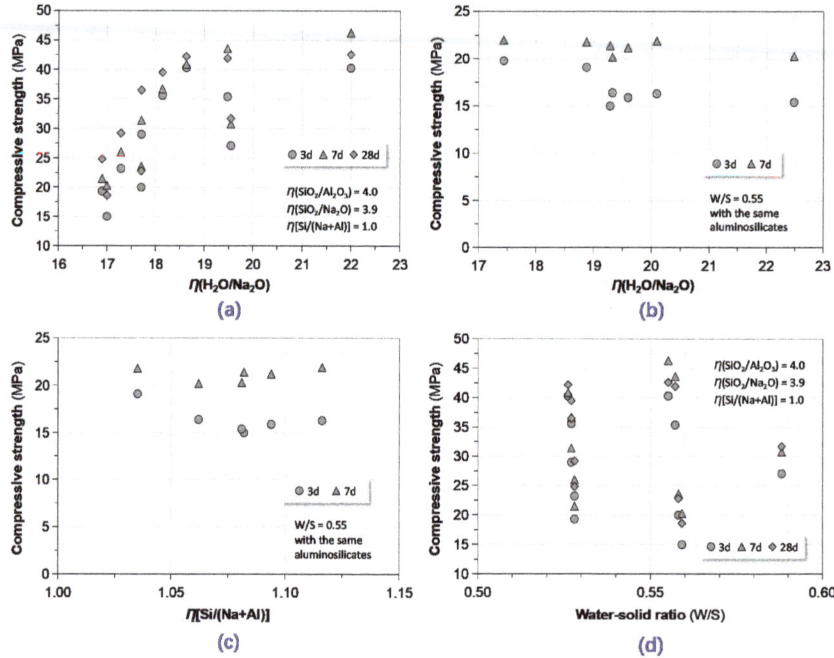

Figure 10. Correlating compressive strength to key molar ratios and W/S: (**a**,**b**) compressive strength vs. $\eta(SiO_2/Na_2O)$; (**c**) compressive strength vs. $\eta[Si/(Na+Al)]$; and (**d**) compressive strength vs. W/S.

3.2.3. Effect of Exposed Conditions

In Figure 11, we compared the standard cured and seawater-immersed GPC's compressive strength. It was easy to find that, although continuous seawater immersion impaired the strength, the loss of 7% was more or less negligible from a viewpoint of practical applications. That is to say, water resistance and anti-corrosion of the GPC mixtures are desirable, even though the exposed condition is harsh. This observation coincided with the results presented by Zhang et al. [27]. The strength reduction of the seawater-immersed specimens could be attributed to dilution of excessive water rather than the corrosion of seawater. Water reservoir also inhibited the polycondensation to some extent [36,51]. Due to excellent impermeability of the GPC, the dilution effect was limited to the surface layer.

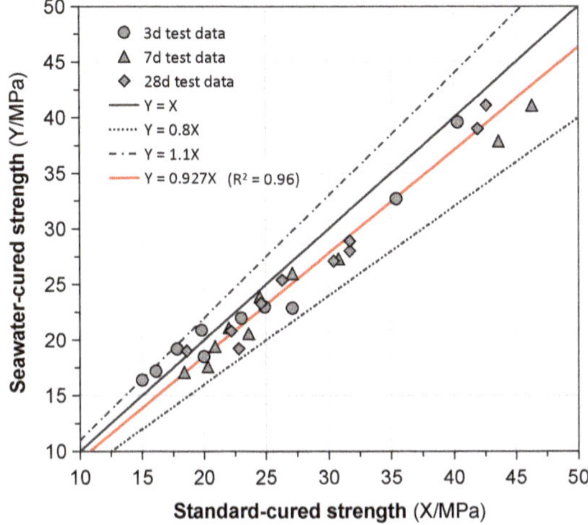

Figure 11. Correlating compressive strength of GPC immersed in artificial seawater to its counterpart stored in standard condition.

3.3. Several Critical Observations

With the exception of all above-discussed factors that affected basic properties of GPC, there are some critical experimental results that are worthwhile to discuss or interpret here.

3.3.1. A Threshold of Total Molar Ratio

In this subsection, with the same W/S, WG and precursor combination, the relationship between setting time/early-age strength and mass of NaOH/η(SiO$_2$/Na$_2$O) was discussed. In this experimental configuration, the only independent variable was the mass of NaOH.

From a phenomenological perspective, it seems as if there was a threshold of the mass of NaOH (30 g) that divided its influence on the strength as well as setting time into two distinct segments (see Figure 12a,b). In the first part, the setting time and strength decreased with an increase of the mass of NaOH, and in the second part, the reverse happened. Because η(SiO$_2$/Na$_2$O) consistently decreased with the increasing addition of NaOH, the correlations between setting time or early-age strength and η(SiO$_2$/Na$_2$O), as shown in Figure 12c,d, highly resembled one another. The threshold of η(SiO$_2$/Na$_2$O) equals to 4.56. Hamidi indicated that the geopolymer paste activated with 12 M NaOH solution obtained the greatest mechanical strength in an experimental study exploring the effect of molarity of NaOH (4–18 M) [55]. In addition, the molarity or content of NaOH in the alkali-activator affected the fluidity of the fresh GPC mixtures and solid particle size [56]. Higher concentration of NaOH solution often leads to better flowability and less particle agglomeration. Similar test results were also observed in this research. The effects of the concentration of alkali metal hydroxide on properties of geopolymer are dependent on the change of zeta potential in the alkaline activator [57].

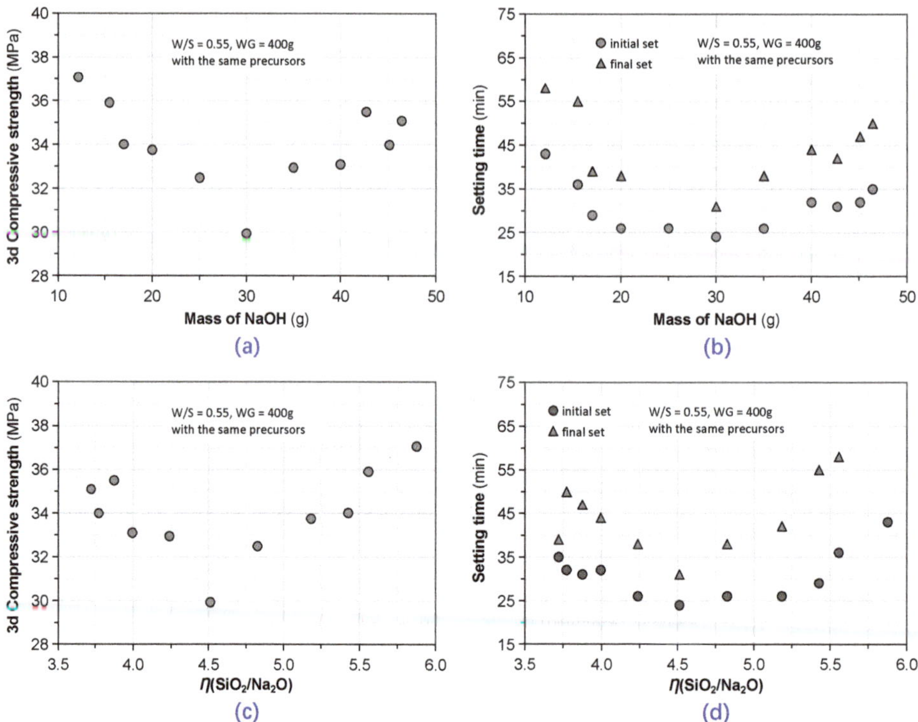

Figure 12. A critical observation on setting time and compressive strength: (**a,b**) compressive strength at the age of 3 days and setting time vs. the mass of NaOH; (**c,d**) compressive strength at the age of 3 days and setting time vs. $\eta(SiO_2/Na_2O)$, respectively.

3.3.2. Effect of OPC Addition

In the binder of geopolymer, OPC, which can cause environmental load such as CO_2 emission, is often not recommended. However, as an important calcium source, the hydration of OPC has two roles: accelerate geopolymerization and produce calcium silicate hydrate (C-S-H) gel [58]. The coexistence of sodium aluminosilicate hydrate (N-A-S-H) and C-S-H will enhance the engineering properties of geopolymer-based materials [59]. In the quadruple precursor, the dosage of OPC should be determined in accordance with the balance between compressive strength and setting behavior of the prepared GPC.

As clearly depicted in Figure 13, setting time and strength properties separately shortened and raised with the increase of OPC replacement ratio in general. Besides, the effects were implicated with the molarity of NaOH in the activator solution; the higher the molarity, the stronger the effects. When the mass ratio of OPC was greater than 10% by total mass of precursor, the GPC's strength tended to decrease, especially the ones activated with higher concentration NaOH solution. Other calcium-rich compounds such as calcium hydroxide or calcium oxide can also be utilized to regulate setting time and mechanical properties; their replacement ratios should be tailored in terms of particular usages [16,22,34].

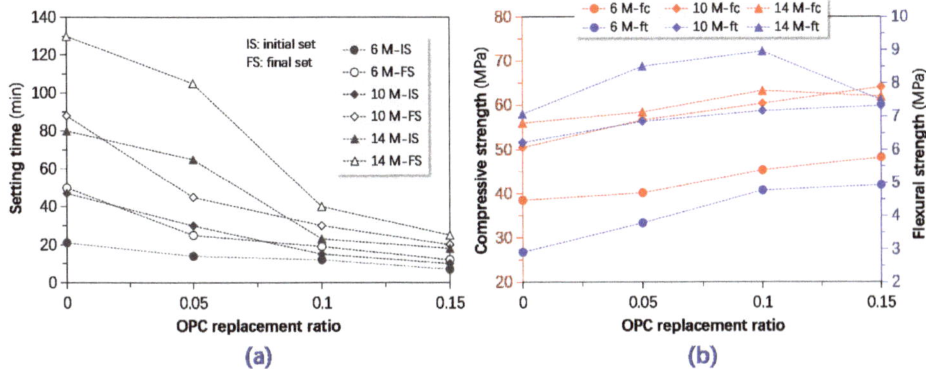

Figure 13. Influences of ordinary Portland cement (OPC) replacement ratio (by mass) on (**a**) setting time and (**b**) compressive and flexural strength.

3.3.3. Effect of GBFS Type

The properties of geopolymer are affected not only by mixture proportions of precursor and activator, but also by type and source of constituent materials. Figure 14 shows the compressive strength of GPC mixtures prepared from four different GBFS with the same recipes. GBFS-B and GBFS-D separately produced the highest and the lowest strength. It appears that this observation was not easy to interpret from key parameters of GBFS such as reactivity index, Al_2O_3 content or total content of Al_2O_3 and SiO_2. The geopolymerization process is complicated and related with the phase morphology of the precursors apart from their combinations and oxide compositions. By the way, the influence of type of GBFS on setting behavior was also ignorable.

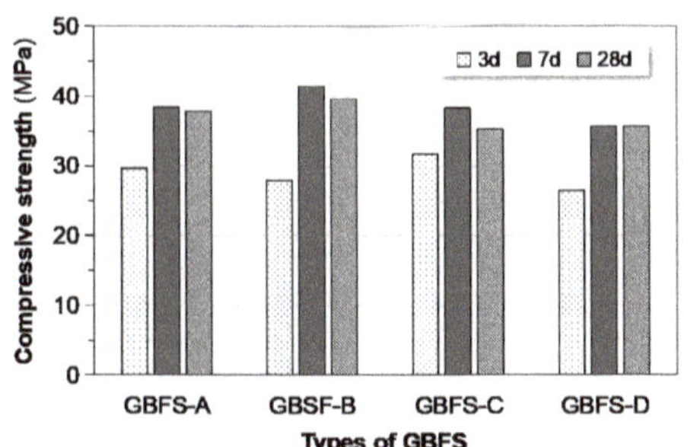

Key parameters of four kinds of GBFS				
Parameters	**GBFS-A**	**GBFS-B**	**GBFS-C**	**GBFS-D**
Reactivity index at 7 days	0.772	0.790	0.780	0.777
Mass ratio of Al_2O_3	0.135	0.125	0.137	0.131
Total mass ratio of Al_2O_3 and SiO_2	0.396	0.379	0.380	0.389

Figure 14. Correlating compressive strength to types of GBFS: several key parameters in terms of reactivity of GBFS are listed alongside.

At this point, the optimized 5 GPC mixtures presented in Table 3 with both practicable setting time (not earlier than 120 min) and desirable compressive strength (not less than 35 MPa) were selected from the investigated 85 mixtures listed in Table 2. In the course of seeking the optimized mixtures, a series of factors affecting the basic properties of GPC were explored and discussed. In particular, an eye was kept on the balance between strength and setting behavior at ambient temperature as well as the usage of quadruple precursors.

Table 3. Adhesive strength of the optimized geopolymer coatings mixtures.

Mixture No.	Precursors (g)				Activators (g)			Additives (g)		Adhesive Strength (MPa)		Compressive Strength (MPa)	
	MK	FA	GBFS-A	OPC	WG	NaOH	W	SP	PPF	7d	28d	7d	28d
OP-1					400.0	85.0	145.8	7.50	0.25	1.763	2.096	43.7	44.0
OP-2					500.0	100.0	110.5	7.50	0.25	1.554	2.613	35.5	35.2
OP-3	75	100	275	50	419.9	24.4	104.0	7.50	0.25	1.646	2.812	35.7	52.5
OP-4					404.3	24.2	110.8	7.50	0.25	2.556	3.398	41.5	52.8
OP-5					400.0	100.0	154.5	7.50	0.25	1.684	1.870	38.5	49.5

3.4. Adhesive Strength

The adhesive strength of the optimized GPC mixtures was measured through a portable tester, as listed in Table 3. All the test results were higher than 1.5 MPa, regardless of the measured age. Previous investigations indicated that the adhesive strength of coatings for most civil applications should not be less than 1.4 to 1.75 MPa, which varied from one testing method to another. According to the provisions in Chinese technical standards, the adhesive property of theses mixtures fully meets the requirement of surface protection. It is easy to understand that higher compressive strength often responds to higher adhesive strength.

3.5. Rapid Chloride Permeability

In order to characterize the compactness of GPC protective layer, as an easily determined indicator, charge passed based on rapid chloride permeability test was explored herein. The preferred mixtures were separately applied to the concrete substrate, and their respective permeability was compared to that of the naked substrate in Figure 15. Apparently, the coated substrate with the GPC obtained improved resistance to deleterious water-soluble ions. The average charge passed of the coated concrete specimens was approximately 30% lower than that of the uncoated ones. The transport mechanism of chemical species in the geopolymer will be further studied.

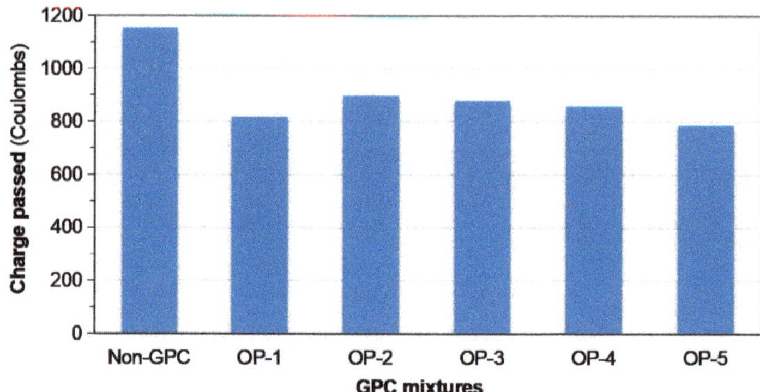

Figure 15. Rapid chloride permeability of coated and uncoated concrete substrate characterized through accelerated charge passed: "Non-GPC" stands for uncoated concrete substrate, and "OP-x" stands for the optimized GPC mixtures listed in Table 3.

3.6. An Outlook on Sustainability

Apart from the engineering properties, the cost, availability and environmental compatibility of GPC are also critical to its potential application. The complexity of aluminosilicate sources, especially aluminosilicate-containing industrial by-products, is a major concern for GPC production. The multi-componential composite precursor is superior to the single precursor, from the perspective of comprehensive utilization of various resources. Due to strong corrosivity and consequent operation safety of alkali metal hydroxides, certain alkaline by-products from chemical industry are more suitable to prepare GPC. Moreover, the end product of GPC possesses preferred life-cycle performance as opposed to organic coating materials. Several authors calculated carbon footprint (or emission factor) and economic cost as well as embodied energy consumption of geopolymer-based product in terms of laboratory or industrial scale [60–63]. Although these analyses referred to different original databases, the results generally indicated that geopolymer is a sustainable alternative material. The preparation of GPC relies on the alkali-activation and the subsequent polycondensation without extra inputs [6]. This process kills two birds with one stone by meeting usage requirements as well as by achieving a cleaner future [64].

4. Conclusions

Apart from as a partial alternative to Portland cement, this paper reports an experimental study on designing, preparing and characterizing a cluster of sustainable GPC that can be potentially utilized to surface the protection of civil infrastructures. Based on the results presented, the following conclusions can be drawn:

- Optimized GPC mixtures with both reasonable setting time and desirable compressive strength were selected from the investigated 85 mixtures. Balancing strength and setting behavior at ambient temperature was viable with the aid of quadruple precursors and mixture design.
- Adhesive strength of preferred GPC mixtures was successfully measured with a portable tester. The test results ranged from 1.5 to 3.4 MPa and fully meet the requirements of surface protection.
- Characterizing GPC's permeability based on a rapid test is feasible and easy to handle. The charge passed of coated concrete specimens with the optimized GPC was 30% lower than that of the uncoated ones.
- Setting time of GPC was positively correlated with $\eta[\text{Si}/(\text{Na}+\text{Al})]$. When the value was greater than 1.1, an abrupt increase of setting time occurred. The extension of setting time with the reduction of $\eta(\text{SiO}_2/\text{Al}_2\text{O}_3)$ was probably ascribed to phase transformation of GPC.
- Compressive strength of GPC was positively affected by the mass contents of GBFS, MK and OPC, and was negatively affected by mass content of FA in the quadruple precursor, respectively. Sustained saline water immersion impaired the strength of GPC to a very limited extent.
- There existed a threshold of mass of NaOH that divided its influence on GPC's strength and setting time into two distinct segments; in the first part, setting time and strength decreased with an increase of NaOH mass, and in the second part, the reverse happened.
- Replacing mass contents of the single components with the total molar ratios of specific oxides or chemical elements as design parameters in the overall GPC system is more reasonable.
- GPC is an innovative material for surface protection and a viable alternative to conventional synthetic polymer coatings for use in civil engineering applications.

Author Contributions: Conceptualization, C.J. and A.W.; methodology, C.J.; software, C.J.; validation, C.J., Z.C. and Z.W.; formal analysis, X.B. and Z.W.; investigation, C.J. and Z.C.; resources, X.B. and Z.W.; data curation, C.J., Z.C. and T.N.; writing—original draft preparation, C.J. and T.N.; writing—review and editing, A.W., T.N.; visualization, C.J. and Z.C.; supervision, A.W., X.B.; project administration, A.W and X.B.; funding acquisition, C.J. All authors have read and agreed to the published version of the manuscript.

Funding: This research was funded by Zhejiang Provincial Natural Science Foundation of China, grant numbers LGG20E090001 & LGF20E080021, and Merit-based Funding Staff of Postdoctoral Science Research Program of Zhejiang Province (no grant number).

Acknowledgments: This work was supported by Zhejiang Provincial Natural Science Foundation of China under Grant Nos. LGG20E090001 & LGF20E080021, Merit-based Funding Staff of Postdoctoral Science Research Program of Zhejiang Province, Research Program of Housing and Urban-Rural Development Department of Zhejiang Province and Major Cultivation Research Program Funds of Zhejiang College of Construction. The authors would like to appreciate these funds and anonymous reviewers.

Conflicts of Interest: The authors declare no conflict of interest.

References

1. Davidovits, J. Chapter 1: Introduction. In *Geopolymer Chemistry & Applications*, 5th ed.; Davidovits, J., Ed.; Institut Géopolymère: Saint-Quentin, France, 2020; pp. 1–18.
2. Zuo, Y.; Nedeljkovic, M.; Ye, G. Pore solution composition of alkali-activated slag/fly ash pastes. *Cem. Concr. Res.* **2019**, *115*, 230–250. [CrossRef]
3. Provis, J.L. Alkali-activated materials. *Cem. Concr. Res.* **2018**, *114*, 40–48. [CrossRef]
4. Geopolymer Institute. Available online: https://www.geopolymer.org/faq/alkali-activated-materials-geopolymers/ (accessed on 28 July 2020).
5. Ng, C.; Alengaram, U.; Wong, L.S.; Mo, K.H.; Jumaat, M.Z.; Ramesh, S. A review on microstructural study and compressive strength of geopolymer mortar, paste and concrete. *Constr. Build. Mater.* **2018**, *186*, 550–576. [CrossRef]
6. Duxson, P.; Provis, J.L.; Lukey, G.C.; Van Deventer, J.S.J. The role of inorganic polymer technology in the development of 'green concrete'. *Cem. Concr. Res.* **2007**, *37*, 1590–1597. [CrossRef]
7. Elsayed, M.E.; Tarek, M.E.; Mostafa, A.S.; Basil, A.E.; Hamdy, A.A. Surface protection of concrete by new protective coating. *Constr. Build. Mater.* **2019**, *220*, 245–252.
8. Jindal, B.B. Investigations on the properties of geopolymer mortar and concrete with mineral admixtures: A review. *Constr. Build. Mater.* **2019**, *227*, 116644. [CrossRef]
9. Amran, Y.M.; Alyousef, R.; Alabduljabbar, H.; El-Zeadani, M. Clean production and properties of geopolymer concrete; A review. *J. Clean. Prod.* **2020**, *251*, 119679. [CrossRef]
10. Lahoti, M.; Tan, K.H.; Yang, E.-H. A critical review of geopolymer properties for structural fire-resistance applications. *Constr. Build. Mater.* **2019**, *221*, 514–526. [CrossRef]
11. Hassan, A.; Arif, M.; Shariq, M. Use of geopolymer concrete for a cleaner and sustainable environment—A review of mechanical properties and microstructure. *J. Clean. Prod.* **2019**, *223*, 704–728. [CrossRef]
12. Yang, T.; Zhang, Z.; Wang, Q.; Wu, Q. ASR potential of nickel slag fine aggregate in blast furnace slag-fly ash geopolymer and Portland cement mortars. *Constr. Build. Mater.* **2020**, *26230*, 119990. [CrossRef]
13. Yang, M.; Paudel, S.R.; Asa, E. Comparison of pore structure in alkali activated fly ash geopolymer and ordinary concrete due to alkali-silica reaction using micro-computed tomography. *Constr. Build. Mater.* **2020**, *236*, 117524. [CrossRef]
14. Singh, N.; Middendorf, B. Geopolymers as an alternative to Portland cement: An overview. *Constr. Build. Mater.* **2020**, *237*, 117455. [CrossRef]
15. Singh, N.; Saxena, S.K.; Kumar, M.; Rai, S. Geopolymer cement: Synthesis, Characterization, Properties and applications. *Mater. Today Proc.* **2019**, *15*, 364–370. [CrossRef]
16. Zhang, P.; Wang, K.; Li, Q.; Wang, J.; Ling, Y. Fabrication and engineering properties of concretes based on geopolymers/alkali-activated binders—A review. *J. Clean. Prod.* **2020**, *258*, 120896. [CrossRef]
17. Zhang, P.; Gao, Z.; Wang, J.; Guo, J.; Ling, Y. Properties of fresh and hardened fly ash/slag based geopolymer concrete: A review. *J. Clean. Prod.* **2020**, *270*, 122389. [CrossRef]
18. Jindal, B.B.; Sharma, R. The effect of nanomaterials on properties of geopolymers derived from industrial by-products: A state-of-the-art review. *Constr. Build. Mater.* **2020**, *25220*, 119028. [CrossRef]
19. Papakonstantinou, C.; Balaguru, P. Geopolymer Chemistry & Applications. In Proceedings of the International SAMPE Symposium and Exhibition, Baltimore, MD, USA, 3–7 June 2007.
20. Delucchi, M.; Barbucci, A.; Cerisola, G. Study of the physico-chemical properties of organic coatings for concrete degradation control. *Constr. Build. Mater.* **1997**, *11*, 365–371. [CrossRef]
21. Hansson, C.; Mammoliti, L.; Hope, B. Corrosion inhibitors in concrete—Part I: The principles. *Cem. Concr. Res.* **1998**, *28*, 1775–1781. [CrossRef]
22. Moutinho, S.; Costa, C.; Andrejkovičová, S.; Mariz, L.; Velosa, A. Assessment of properties of metakaolin-based geopolymers applied in the conservation of tile facades. *Constr. Build. Mater.* **2020**, *259*, 119759. [CrossRef]
23. Sadowski, Ł.; Hoła, J.; Żak, A.; Chowaniec, A. Microstructural and mechanical assessment of the causes of failure of floors made of polyurethane-cement composites. *Compos. Struct.* **2020**, *238*, 112002. [CrossRef]
24. Franzoni, E.; Pigino, B.; Pistolesi, C. Ethyl silicate for surface protection of concrete: Performance in comparison with other inorganic surface treatments. *Cem. Concr. Compos.* **2013**, *44*, 69–76. [CrossRef]

25. Singh, N.; Mukesh, K.; Sarita, R. Geopolymer cement and concrete: Properties. *Mater Today*: Proceedings in press, corrected proof. Available online: https://www.sciencedirect.com/science/article/pii/S2214785320331357/ (accessed on 5 May 2020).

26. Lahoti, M.; Wijaya, S.F.; Tan, K.H.; Yang, E.H. Tailoring sodium-based fly ash geopolymers with variegated thermal performance. *Cem. Concr. Compo.* **2020**, *107*, 103507. [CrossRef]

27. Zhang, Z.; Yao, X.; Zhu, H. Potential application of geopolymers as protection coatings for marine concreteI. Basic properties. *Appl. Clay Sci.* **2010**, *49*, 1–6. [CrossRef]

28. Aguirre, G.; María, A.; Rafael, A. A novel geopolymer application: Coatings to protect reinforced concrete against corrosion. *Appl. Clay. Sci.* **2017**, *135*, 437–446. [CrossRef]

29. Wiyono, D.; Antoni; Hardjito, D. Improving the Durability of Pozzolan Concrete Using Alkaline Solution and Geopolymer Coating. *Procedia Eng.* **2015**, *125*, 747–753. [CrossRef]

30. Lv, X.; Wang, K.; He, Y.; Cui, X. A green drying powder inorganic coating based on geopolymer technology. *Constr. Build. Mater.* **2019**, *214*, 441–448. [CrossRef]

31. Wang, Y.; Zhao, J. Preliminary study on decanoic/palmitic eutectic mixture modified silica fume geopolymer-based coating for flame retardant plywood. *Constr. Build. Mater.* **2018**, *189*, 1–7. [CrossRef]

32. Chindaprasirt, P.; Phoo-Ngernkham, T.; Hanjitsuwan, S.; Horpibulsuk, S.; Poowancum, A.; Injorhor, B. Effect of calcium-rich compounds on setting time and strength development of alkali-activated fly ash cured at ambient temperature. *Case Stud. Constr. Mater.* **2018**, *9*, e00198. [CrossRef]

33. Elkem. Elkem Materials Mixture Analyser—EMMA User Guide. Available online: https://www.elkem.com/silicon-materials/construction/concrete/ (accessed on 28 July 2020).

34. Aïtcin, P.C. *High-Performance Concrete*; E & FN SPON: London, UK, 1998; pp. 147–163.

35. Mehta, P.K.; Paulo, J. *Concrete: Microstructiure, Properties and Materials*; McGraw-Hill: New York, NY, USA, 2006; pp. 37–45.

36. Phoo-Ngernkham, T.; Maegawa, A.; Mishima, N.; Hatanaka, S.; Chindaprasirt, P. Effects of sodium hydroxide and sodium silicate solutions on compressive and shear bond strengths of FA–GBFS geopolymer. *Constr. Build. Mater.* **2015**, *91*, 1–8. [CrossRef]

37. Tho-In, T.; Sata, V.; Boonserm, K.; Chindaprasirt, P. Compressive strength and microstructure analysis of geopolymer paste using waste glass powder and fly ash. *J. Clean. Prod.* **2018**, *172*, 2892–2898. [CrossRef]

38. Amran, Y.M.; Soto, M.G.; Alyousef, R.; El-Zeadani, M.; Aune, V. Performance investigation of high-proportion Saudi-fly-ash-based concrete. *Resu. Eng.* **2020**, *6*, 100118. [CrossRef]

39. Davidovits, J. Chapter 3: Macromolecular structure of natural silicates and alumino-silicates. In *Geopolymer chemistry & applications*, 5th ed.; Davidovits, J., Ed.; Institut Géopolymère: Saint-Quentin, France, 2020; pp. 55–68.

40. Davidovits, J. High-alkali cements for 21st century concretes. *Spec. Publ.* **1994**, *144*, 383–398.

41. Rangan, B. Fly ash-based geopolymer concrete for environmental protection. *Indian Concr. J.* **2014**, *88*, 41–59.

42. Aliabdo, A.A.; Elmoaty, A.E.M.A.; Salem, H.A. Effect of water addition, plasticizer and alkaline solution constitution on fly ash based geopolymer concrete performance. *Constr. Build. Mater.* **2016**, *121*, 694–703. [CrossRef]

43. Xiao, R.; Ma, Y.; Jiang, X.; Zhang, M.; Zhang, Y.; Wang, Y.; Huang, B.; He, Q. Strength, microstructure, efflorescence behavior and environmental impacts of waste glass geopolymers cured at ambient temperature. *J. Clean. Prod.* **2020**, *252*, 119610. [CrossRef]

44. ASTM International. *Standard Test Methods for Time of Setting of Hydraulic Cement by Vicat Needle (ASTM C191-19)*; American Society for Testing Materials: West Conshohocken, PA, USA, 2019.

45. Rifaai, Y.; Yahia, A.; Mostafa, A.; Aggoun, S.; Kadri, E.-H. Rheology of fly ash-based geopolymer: Effect of NaOH concentration. *Constr. Build. Mater.* **2019**, *223*, 583–594. [CrossRef]

46. ASTM International. *Standard Test Method for Electrical Indication of Concrete's Ability to Resist Chloride Ion Penetration (ASTM C1202-19)*; American Society for Testing Materials: West Conshohocken, PA, USA, 2019.

47. Li, N.; Shi, C.; Zhang, Z. Understanding the roles of activators towards setting and hardening control of alkali-activated slag cement. *Compos. Part B Eng.* **2019**, *171*, 34–45. [CrossRef]

48. Arnoult, M.; Perronnet, M.; Autef, A.; Rossignol, S. How to control the geopolymer setting time with the alkaline silicate solution. *J. Non-Crystalline Solids* **2018**, *495*, 59–66. [CrossRef]

49. Ozer, I.; Soyer-Uzun, S. Relations between the structural characteristics and compressive strength in metakaolin based geopolymers with different molar Si/Al ratios. *Ceram. Int.* **2015**, *41*, 10192–10198. [CrossRef]

50. Chen, S.; Wu, C.; Yan, D. Binder-scale creep behavior of metakaolin-based geopolymer. *Cem. Concr. Res.* **2019**, *124*, 105810. [CrossRef]

51. Wang, Y.-S.; Alrefaei, Y.; Dai, J.-G. Silico-Aluminophosphate and Alkali-Aluminosilicate Geopolymers: A Comparative Review. *Front. Mater.* **2019**, *6*, 233–247. [CrossRef]

52. Cheng, H.; Lin, K.L.; Cui, R.; Hwang, C.L.; Cheng, T. The effects of SiO_2/Na_2O molar ratio on the characteristics of alkali-activated waste catalyst-metakaolin based geopolymers. *Constr. Build. Mater.* **2015**, *95*, 710–720. [CrossRef]

53. Gao, K.; Lin, K.L.; Wang, D.Y.; Hwang, C.L.; Cheng, T.W. Effects SiO_2/Na_2O molar ratio on mechanical properties and the microstructure of nano-SiO_2 metakaolin-based geopolymers. *Constr. Build. Mater.* **2014**, *53*, 503–510. [CrossRef]

54. Juengsuwattananon, K.; Winnefeld, F.; Chindaprasirt, P.; Pimraksa, K. Correlation between initial SiO_2/Al_2O_3, Na_2O/Al_2O_3, Na_2O/SiO_2 and H_2O/Na_2O ratios on phase and microstructure of reaction products of metakaolin-rice husk ash geopolymer. *Constr. Build. Mater.* **2019**, *226*, 406–417. [CrossRef]

55. Hamidi, R.M.; Man, Z.; Azizli, K.A. Concentration of NaOH and the Effect on the Properties of Fly Ash Based Geopolymer. *Procedia Eng.* **2016**, *148*, 189–193. [CrossRef]

56. Liu, Y.; Shi, C.; Zhang, Z.; Li, N. An overview on the reuse of waste glasses in alkali-activated materials. *Resour. Conserv. Recycl.* **2019**, *144*, 297–309. [CrossRef]

57. Zhang, D.-W.; Wang, D.; Liu, Z.; Xie, F.-Z. Rheology, agglomerate structure, and particle shape of fresh geopolymer pastes with different NaOH activators content. *Constr. Build. Mater.* **2018**, *187*, 674–680. [CrossRef]

58. Wang, R.; Jingsong, W.; Teng, D.; Gaoshang, O. Structural and mechanical properties of geopolymers made of aluminosilicate powder with different SiO_2/Al_2O_3 ratio: Molecular dynamics simulation and microstructural experimental study. *Constr. Build. Mater.* **2020**, *240*, 117935. [CrossRef]

59. Lodeiro, I.G.; Palomo, A.; Fernandez-Jiménez, A.; Macphee, D. Compatibility studies between N-A-S-H and C-A-S-H gels. Study in the ternary diagram Na_2O–CaO–Al_2O_3–SiO_2–H_2O. *Cem. Concr. Res.* **2011**, *41*, 923–931. [CrossRef]

60. Pan, X.; Shi, Z.; Shi, C.; Ling, T.-C.; Li, N. A review on concrete surface treatment Part I: Types and mechanisms. *Constr. Build. Mater.* **2017**, *132*, 578–590. [CrossRef]

61. Jiang, L.; Xue, X.; Zhang, W.; Yang, J.; Zhang, H.; Li, Y.; Zhang, R.; Zhang, Z.; Xu, L.; Qu, J.; et al. The investigation of factors affecting the water impermeability of inorganic sodium silicate-based concrete sealers. *Constr. Build. Mater.* **2015**, *93*, 729–736. [CrossRef]

62. Ohno, M.; Li, V.C. An integrated design method of Engineered Geopolymer Composite. *Cem. Concr. Compos.* **2018**, *88*, 73–85. [CrossRef]

63. Fahim, H.; Ghasan, R.; Jahangir, M. Geopolymer mortars as sustainable repair material: A comprehensive review. *Renew. Sust. Engy. Rev.* **2017**, *80*, 54–74.

64. Turner, L.K.; Collins, F. Carbon dioxide equivalent (CO_2-e) emissions: A comparison between geopolymer and OPC cement concrete. *Constr. Build. Mater.* **2013**, *43*, 125–130. [CrossRef]

Article

Corrosion Resistance of Concrete Reinforced by Zinc Phosphate Pretreated Steel Fiber in the Presence of Chloride Ions

Xingke Zhao [1], Runqing Liu [1,*], Wenhan Qi [2] and Yuanquan Yang [1]

[1] School of Materials Science and Engineering, Shenyang Ligong University, Shenyang 110159, China; zczczc2000@126.com (X.Z.); aquarius0109@163.com (Y.Y.)

[2] Technology R & D Department, Zhejiang Dadongwu Group Co., Ltd., Huzhou 313000, China; an77qt@163.com

* Correspondence: liurunqing@sylu.edu.cn

Received: 12 July 2020; Accepted: 14 August 2020; Published: 17 August 2020

Abstract: This paper aims to provide new insight into a method to improve the chloride ion corrosion resistance of steel fiber reinforced concrete. The steel fiber was pretreated by zinc phosphate before the preparation of the fiber reinforced concrete. Interfacial bond strength, micro-hardness and micro-morphology properties were respectively analyzed in the steel fiber reinforced concrete before and after the chloride corrosion cycle test. The results show that the chloride ion corrosion resistance of the steel fiber was enhanced by zinc phosphate treatment. Compared to plain steel fiber reinforced concrete under chloride ion corrosion, the interfacial bond strength of the concrete prepared by steel fiber with phosphating treatment increased by 15.4%. The thickness of the interface layer between the pretreated steel fiber and cement matrix was reduced by 50%. The micro-hardness of the weakest point in the interface area increased by 54.2%. The micro-morphology of the interface area was almost unchanged before and after the corrosion. The steel fiber reinforced concrete modified by zinc phosphate can not only maintain the stability of the microstructure when corroded by chloride ion but also presents good bearing capacity.

Keywords: steel fiber reinforced concrete; zinc phosphate; chloride ion corrosion resistance; microstructure

---

1. Introduction

Steel fiber reinforced concrete (SFRC) is a new type of multiphase composite material which is formed by mixing short steel fibers with random distribution in plain concrete [1–3]. For concrete, steel fiber can prevent cracks and increase the toughness of the concrete [4,5]. However, chloride ion, as a corrosive medium in the marine environment, can easily cause corrosion of steel fiber, which leads to loss of bearing capacity and toughness of steel fiber reinforced concrete members [5–7]. In the process of preparation of steel fiber reinforced concrete, a layer of interface phase with a thickness of around 50–100 μm will be formed when the steel fiber is wrapped by the substrate. Due to the particularity of its structure, the bond strength, fracture toughness, and hardness of the interface are all weak points in SFRC [8]. Unfortunately, the corrosion of steel fiber under chloride ion attack will seriously affect the performance of the interface area, resulting in the reduction of the performance of steel fiber reinforced concrete. Therefore, the corrosion protection of steel fiber reinforced concrete should not be ignored [9–14].

There are many factors affecting the corrosion of steel fiber in concrete, such as surrounding environment, selection of material as ingredients for concrete [15], steel fiber condition, exposure time, and cracks resulting from the brittle nature of the concrete [16,17]. Thus, careful selection of materials, suitable mixing proportion design for concrete, and protection for the steel fiber are beneficial for the anti-corrosion of steel fiber. Previous studies [18] have found that a high water/binder ratio for concrete under a corrosive environment resulted in accelerated corrosion of steel fiber. Concrete with low water/binder ratio and permeability can resist chloride penetration into the concrete matrix and provide a barrier against oxygen entry and, therefore, the time of corrosion initiation for the steel fiber can be extended. In addition, waste by-products like fly ash, silica fume, and granulated blast furnace slag were used to partially replace the binder, which decreased the permeability of the concrete and thus the steel fiber was protected from corrosion [19–21]. Research works also concentrate on methods to increase the crack resistance of concrete structures and therefore lessen the attacks of aggressive agents such as chloride ions, including the addition of steel lining and/or pre-stressing of the concrete. Moreover, cracks of the concrete are repaired by epoxy resins and/or urethanes [5]. All these methods have their limitations. Careful selection of materials for concrete and suitable mixing design mostly work on dormant cracks and thus the steel fibers are protected, but when the concrete is continuously subjected to dynamic cyclic loads, as in an offshore environment, it fails to protect the steel fiber. The additional steel lining and pre-stressing for the concrete significantly increase the construction cost as well. Moreover, for steel fiber reinforced concrete, the fibers are dispersed in the entire matrix, located close to the surface and with less concrete matrix cover. Therefore, corrosion of steel fiber cannot be avoided and effective methods are required in order to solve this problem.

Recently, researchers found that phosphating was an effective and important process applied to steels which has been successfully used in automotive industries due to easy operation and low cost [22]. The phosphating method (phosphate coats) can improve the corrosion resistance and lubrication properties of steel fibers [23] as well as steel rebars [24,25]. Among phosphate coats, zinc phosphate is the most popular coating, due to its capacity for anodic protection of Zn^{2+} and barrier protection of the existing phosphate film, and, therefore, phosphate coatings have attracted more attention in recent years. Phosphating as an established method is used in various industries and a majority of the phosphating is applied in automotive industries. However, zinc phosphating of steel fiber for concrete has not been sufficiently investigated. The strength of the interfacial bond between the zinc phosphate treated steel fiber and the concrete matrix and the ways in which microstructures of the concrete matrix are affected by the incorporation of the pretreated steel fiber are not clear. These are the key parameters that affect the durability of the concrete [26].

This work thus aims to investigate the interfacial bond strength and microstructure of steel fiber reinforced concrete by zinc phosphate treatment on the steel fiber. The steel fiber was pretreated by zinc phosphate before the preparation of the fiber reinforced concrete. Interfacial bond strength, micro-hardness, and micro-morphology properties were respectively analyzed in the prepared steel fiber reinforced concrete before and after the chloride corrosion cycle test. The concrete with modified steel fiber could be more suitable to be the first choice material of load-bearing structures in severe corrosive environments so as to greatly expand the scope of use of steel fiber reinforced concrete. It can provide safer and more reliable load-bearing materials for future engineering construction in various corrosive environments.

2. Experimental Methods

2.1. Materials

CEM 42.5 Portland cement produced by Ji Dong Cement Co., Ltd. of Anshan, Liaoning, China, was used in the experiment. Grade II fly ash was produced by Shenhai Thermal Power Plant, Shenyang, China. The chemical composition and physical properties of Portland cement and fly ash are presented in Table 1. The steel fiber produced by Shanghai Shiweike Industry (Shanghai, China) and Trade Co., Ltd. (Shanghai, China) has a fiber length of 40 mm, a fiber length diameter ratio of 69.3, and a tensile strength of ≥600 MPa. River sand (medium sand) with fineness modulus of 2.60 and crushed stone with particle size of 5–25 mm were also used when making the concrete specimens.

Table 1. Chemical composition and physical properties of Portland cement and fly ash.

Materials	CaO	SiO$_2$	Al$_2$O$_3$	Fe$_2$O$_3$	MgO	SO$_3$	R$_2$O	Loss	d$_{mean}$/μm	BET/(m^2/g)
CEM 42.5	64.47	22.68	5.81	4.47	1.74	2.65	0.51	1.49	13.7	3.14
Fly ash	4.35	59.95	26.78	1.53	2.30	1.46	2	1.63	12.5	7.81

d$_{mean}$ is average particle size. BET is specific suface area mesured by Brunner–Emmet–Teller method.

2.2. Methods

Phosphating solution preparation: 8.0 g ZnO powder and moderate amounts (about 15 g) of distilled water were mixed; then, 20 mL of 85% H$_3$PO$_4$ was added as well as 1.0 g of citric acid, 0.2 g of Ca(NO$_3$)$_2$, and 0.2 g of Zn(NO$_3$)$_2$. The composition of this mixture after reactants' reaction mainly comprised of Zn$_3$(PO$_4$)$_2$ and minor amounts of Ca(NO$_3$)$_2$ and Zn(NO$_3$)$_2$, which could also be seen from the XRD identification (Figure 1c) of the coating of the steel fiber that was modified by the as-prepared zinc phosphate solution. The addition of minor amounts of Ca(NO$_3$)$_2$ and Zn(NO$_3$)$_2$ here was to improve the formation of the zinc phosphate coating layer for the steel fiber. Finally, additional distilled water was added in order to obtain 1 L of the phosphating solution.

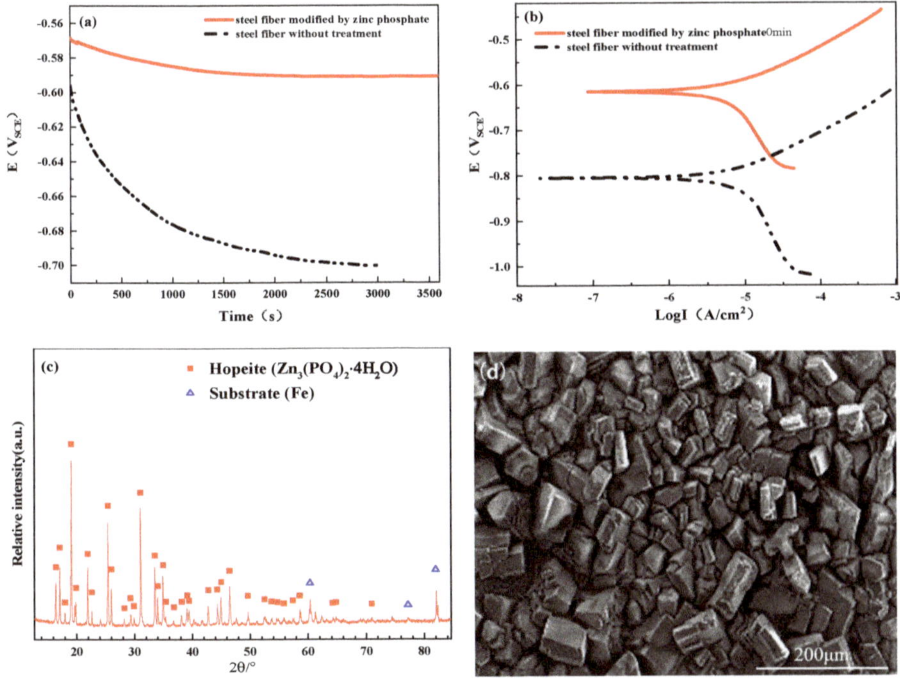

Figure 1. Performance index of the steel fiber modified by zinc phosphate at 75 °C for 20 min: (**a**) ppen circuit voltage, (**b**) polarization curve, (**c**) XRD identification, and (**d**) SEM morphology.

During the treatment of steel fiber by zinc phosphate solution, the oil and rust on the steel fiber was firstly removed, and it was then treated with the zinc phosphate solution. Each process was performed separately and in turn. In particular, in the process regarding the oil and rust removal from the steel fiber, the steel fiber was treated in accordance with the following procedures: (a) polished, (b) soaked in 10% NaOH solution at 50 °C in water-bath for 10 min, (c) soaked in acetone for 2 min and then distilled water for another 2 min before being dried, (d) soaked in 25% H$_2$SO$_4$ solution at 60 °C in water bath for 5 min, (e) repeat the third step, (f) soaked in 25% H$_3$PO$_4$ at ambient temperature for 5 min, and (g) soaked in 6% citric acid solution for 2 min before being dried. According to the results of previous experiments, the steel fiber was phosphatized at 75 °C for 20 min to obtain better performance. The performance index of the steel fiber modified by zinc phosphate at 75 °C for 20 min is shown in Figure 1. The open circuit voltage and polarization curve (Figure 1a,b) indicates that steel fiber modified by zinc phosphate presents better corrosion resistance than the non-treated one. Moreover, the coated layer of the steel fiber modified by zinc phosphate mainly contains a hopeite phase ($(Zn_3(PO_4)_2 \cdot 4H_2O)$, Figure 1c) with irregular structures (Figure 1d), as well as an Fe phase (substrate).

Electrochemical tests were performed on the plain steel fiber and zinc phosphate modified steel fiber. During the electrochemical test, the samples were suspended in 3.5 wt.% NaCl solution at a scan rate of 0.01 mV/s from $-0.2 \sim 0.67$ V using an EG and G237 electrochemical workstation. The corrosion potential (E_{corr}), corrosion currency density (I_{corr}), polarization resistance (R_p), corrosion rate, porosity of the coating layer (P), and corrosion protection efficiency (P_e) for the prepared steel fiber phosphatized at 75 °C for 20 min and that of plain steel fiber were determined by the Tafel extrapolation method [27] and are shown in Table 2.

Table 2. The results of potentiodynamic corrosion tests.

No.	E_{corr} (mV)	I_{corr} ($\mu A/cm^2$)	$R_p \times 10^3$ ($\Omega \cdot cm^2$)	Corrosion Rate (g/m^2h)	P (%)	Pe (%)
SF[a]	-805 ± 5.3	31.998 ± 1.2	3.06 ± 0.87	0.333	-	-
ZPP-SF[b]	-641 ± 5.2	8.721 ± 1.0	60.2 ± 3.2	0.091	1.53	72.7

SF[a] is plain steel fiber. ZPP-SF[b] is zinc phosphate modified steel fiber.

Concrete specimens with a water to binder ratio of 0.46 and 1% volume of steel fiber (in relation to the total volume of concrete) were prepared according to the "Test method for steel fiber reinforced concrete (CECS 1389, China standard)". The detailed mixing proportions are presented in Table 3.

Table 3. Mixing proportions of the concrete.

Mixing Proportions of the Concrete (kg/m^3)					Volume of Steel Fiber
Portland Cement	Fly Ash	Water	Sand	Gravel	
526	130	305	808	987	1%

The concrete specimens with sizes of 100 mm × 100 mm × 400 mm were treated 30 times with chloride ion solution cycles. For each cycle, the specimens were immersed in 6% NaCl solution for 12 h at 20 °C and then dried for another 12 h. After the chloride ion corrosion test, the concrete specimens were tested for various tests. Microstructures of the corresponding specimens were observed by SEM (S-4800, Hitachi, Tokyo, Japan).

Crushed concrete specimens (5 mm in diameter) that contained interfacial transition zones between the steel fiber and cement matrix were chosen for micro-hardness measurement (FM-700 micro-hardness tester, Tokyo, Japan, FUTURE-TECH). Each sample was polished to obtain a smooth fracture surface before the micro-hardness measurement, to ensure its accuracy. Micro-hardness data were recorded from the substrate of the cement matrix at various points around 100 μm away.

The interfacial bond strength between the steel fiber and the cement substrate was measured according to the "Test method for steel fiber reinforced concrete (CECS 1389, China standard)". For the micro-hardness and interfacial bond strength tests, three samples (with the same type of steel fiber and under the same environment) were tested.

3. Results and Discussion

3.1. Bond Strength

Two forces exist in the process of pulling out steel fiber from the cement matrix. Cohesive force plays a key role between the fiber and cement matrix before the arrival of a peak load, and when the fiber de-bonds from the cement matrix, friction between the fiber and the cement matrix becomes the main working factor. A diagram showing the steel fiber pull-out from the cement matrix is shown in Figure 2.

Figure 2. Pull-out test for the steel fiber from the cement matrix.

Generally, the load displacement curve produced by fiber pull-out is divided into three stages. In the first stage, the interface between the steel fiber and concrete is elastic, which is referred to as linear loading. In the second stage, the steel fiber slips from the concrete and gradually loses its stability, which is referred to as non-linear loading. In the third stage, the steel fiber is de-bonded from the concrete matrix, and the steel fiber is pulled out of the concrete matrix, which is referred to as non-linear unloading. The mathematical expression of bond strength between steel fiber and concrete matrix is given in Equation (1).

$$f_{fu} = \frac{F_{fu}}{n u_f l_{fe}} \tag{1}$$

where f_{fu} denotes the interfacial bond strength of steel fiber and concrete (MPa), F_{fu} is the maximum load when steel fiber is pulled out (N), n is the embedded quantity of steel fiber, u_f is the perimeter of the steel fiber cross-section (mm), and l_{fe} is the embedded length of the steel fiber (mm).

The toughening effect of steel fiber on concrete is mainly due to the large amount of work required when the fiber is pulled out from the matrix. Therefore, the bond strength between the steel fiber and matrix determines its toughening effect on the concrete. The mechanical properties of the steel fiber in the test were obtained by calculation, as presented in Table 4.

Table 4 shows that the corrosion environment has a great influence on the interfacial bond performance of plain steel fiber (without treatment) reinforced concrete. The average bond strength of the interface between steel fiber and concrete obviously decreased after corrosion, as well as the work carried out when the fiber was pulled out. The average bond strength of the interface between the steel fiber and cement was 15.4% lower than that of phosphatized steel fiber, and 15.0% less work was required when the fiber was pulled out. It can be seen that the corrosion resistance of steel fiber has been greatly improved after phosphating treatment. This can ensure improvement in the corrosion resistance and bond strength of the interface after initiation of corrosion.

Table 4. Bond strength of the interface between the steel fiber and concrete matrix.

Environment	Type of Fiber	Embedded Depth of the Steel Fiber/mm	Average Value of Maximum Pull-Out Load/N	Average Interfacial Bond Strength/MPa	Work Done When the Fiber Is Pulled Out/N·m
Corroded	SF	16	1218.56	6.35	0.60
	ZPP-SF	16	1407.81	7.33	0.69
Uncorroded	SF	16	1243.03	6.47	0.61
	ZPP-SF	16	1323.76	6.89	0.65

SF is plain steel fiber. ZPP-SF is zinc phosphate modified steel fiber.

To further study the influence of phosphatized steel fiber on the interfacial bond strength of concrete specimens under different corrosion environments, the relationship between the load and displacement of the steel fiber reinforced concrete specimens in different corrosion environments before and after phosphating was plotted by recording and taking the average value of each type of data (Figure 3).

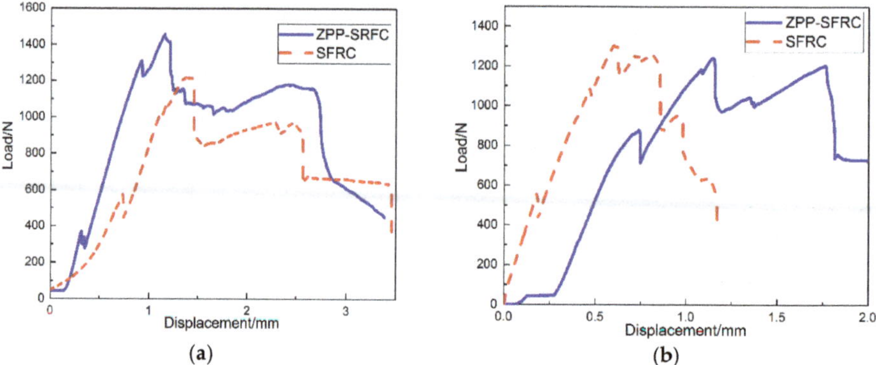

Figure 3. The relationship between load and displacement of the samples under different environments in the pull-out test: (**a**) NaCl solution and (**b**) without corrosion.

By comparing the load displacement curve of steel fiber reinforced concrete modified by zinc phosphate and plain steel fiber reinforced concrete, it can be seen that the ultimate load strength of steel fiber reinforced concrete modified by zinc phosphate was significantly higher than that of plain steel fiber reinforced concrete in a chloride corrosion environment. The difference in ultimate load between steel fiber reinforced concrete modified by zinc phosphate and plain steel fiber reinforced concrete was not significant in the environment without corrosion. The displacement of steel fiber reinforced concrete modified by zinc phosphate, however, shifted around 0.6 mm towards the right compared to plain steel fiber reinforced concrete, as well as withstanding higher load. This demonstrates that the steel fiber modified by zinc phosphate has good interfacial adhesion.

3.2. Micro-Hardness

For steel fiber concrete, the interface layer is a contact layer between the fibers and the concrete matrix. The micro-hardness value of the interface layer is an intuitive response to the performance of the interface area. It can reflect the interface effect between the steel fiber and concrete matrix and the size of the interface area. The sample corroded by NaCl solution was polished and leveled and then observed by a micro-hardness tester. The micro-hardness values from the tests are presented in

Figure 4. The comparison of the interface between steel fiber and concrete modified by zinc phosphate in a chloride corrosion environment and non-corrosion environment is shown in Figure 5.

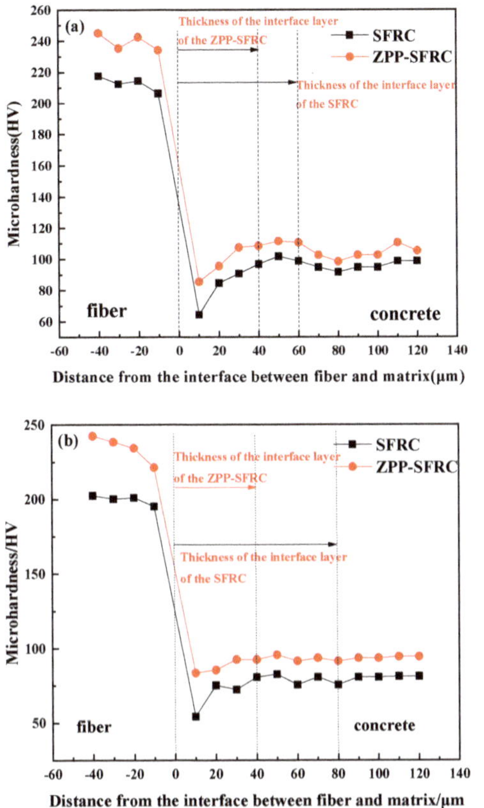

Figure 4. Micro-hardness of the interface layer between ZPP-SFRC (zinc phosphate modified steel fiber reinforced concrete) and SFRC (steel fiber reinforced concrete) under different environments: (**a**) without corrosion and (**b**) with NaCl corrosion.

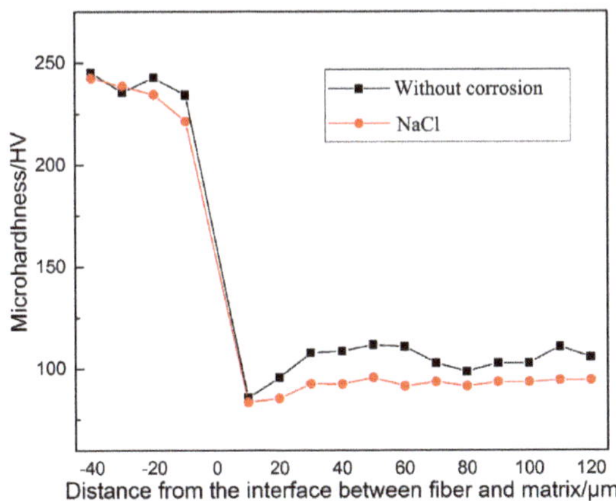

Figure 5. Micro-hardness of interface layer between zinc phosphate modified fiber and concrete matrix.

It can be seen from Figure 4 that phosphating treatment decreases the thickness of the interface layer between the steel fiber and concrete by 20 μm, as well as increasing its micro-hardness by around 20 HV in both corroded and uncorroded environments. With regards to the plain steel fiber reinforced concrete, the presence of a corroded environment increases the thickness of the interface layer by 20 μm and decreases the micro-hardness by 15 HV, compared to that under the uncorroded environment. However, the thickness of the interface layer of the steel fiber reinforced concrete modified by zinc phosphate did not change much and remained at around 40 μm. In the process of steel fiber extending to the concrete matrix, the micro-hardness value decreased initially and then increased and finally tended to be stable. There was a minimum value of micro-hardness curve, and the weak valley was the most vulnerable part of the structure. Compared with the micro-hardness value of the weakest point of the interface, the micro-hardness of the weakest point of the sample increased, regardless of whether it was corroded by NaCl solution or not after phosphating, especially for the sample after corrosion. The micro-hardness value of the weakest point of the interface of plain steel fiber reinforced concrete after NaCl corrosion was 54.257 MPa, while this value was 83.654 MPa for the steel fiber reinforced concrete modified by zinc phosphate, which increased by 54.2%. The results show that the interface structure between the steel fiber and concrete matrix can be effectively improved by zinc phosphate modification. It can decrease the thickness of the interface layer, improve the hardness of the interface layer, reduce the influence of chloride corrosion on the interface layer of steel fiber reinforced concrete, and finally improve the interface performance.

It was found that the micro-hardness of the interface between the steel fiber reinforced concrete modified by zinc phosphate and that corroded by NaCl solution are similar. The change in the range of micro-hardness was around 1%. This shows that the corrosion of the NaCl solution has little effect on the interface layer between the steel fiber and concrete matrix modified by zinc phosphate, which is due to the good corrosion resistance of the zinc phosphate layer, which can resist chloride corrosion and bond well with the concrete matrix.

3.3. SEM-EDS Analysis

The microstructure of the concrete matrix after the pull-out of the fiber is presented in Figure 6. The interface area adjacent to the steel fiber for the SFRC (Figure 6a) was slightly different from that of ZPP-SFRC (Figure 6b), which tends to be a porous structure with small pores rather than a simple, dense, and smooth structure with calcium silicate hydrate (C–S–H) gel. This could be attributed to the different properties of the plain steel fiber and zinc phosphate modified fiber in the vicinity of their surfaces. In the process of hydration, the cement hydrates well as a result of the fact that more water was absorbed by the zinc phosphate modified fiber in the vicinity of their surfaces, thus forming a denser and smoother structure (Figure 6b). After NaCl corrosion, the structure of the steel fiber concrete without phosphating treatment has quite a high number of large pores (Figure 6c). However, the micro-morphology of steel fiber reinforced concrete modified by zinc phosphate did not change much and only a low quantity of small pores appeared (Figure 6d). This may be due to the excellent corrosion resistance and surface roughness of zinc phosphate modified steel fiber. The excellent corrosion resistance of the zinc phosphate modified steel fiber could prevent the growth of rust crystals on its surface, which could mitigate the expansion by the rust crystal formation; thus, the cracks on the concrete proceed very slowly and the corrosion resistance of the steel fiber concrete is improved.

A mercury pressure intrusion test can be used to analyze the pore structure of concrete. Figure 7 is a comparison of mercury intrusion before and after NaCl solution corrosion between zinc phosphate modified steel fiber concrete and plain steel fiber concrete. After NaCl corrosion, the cumulative mercury intrusion of steel fiber reinforced concrete is higher than that of zinc phosphate modified steel fiber. It shows that the porosity of steel fiber reinforced concrete is higher than that of zinc phosphate modified steel fiber.

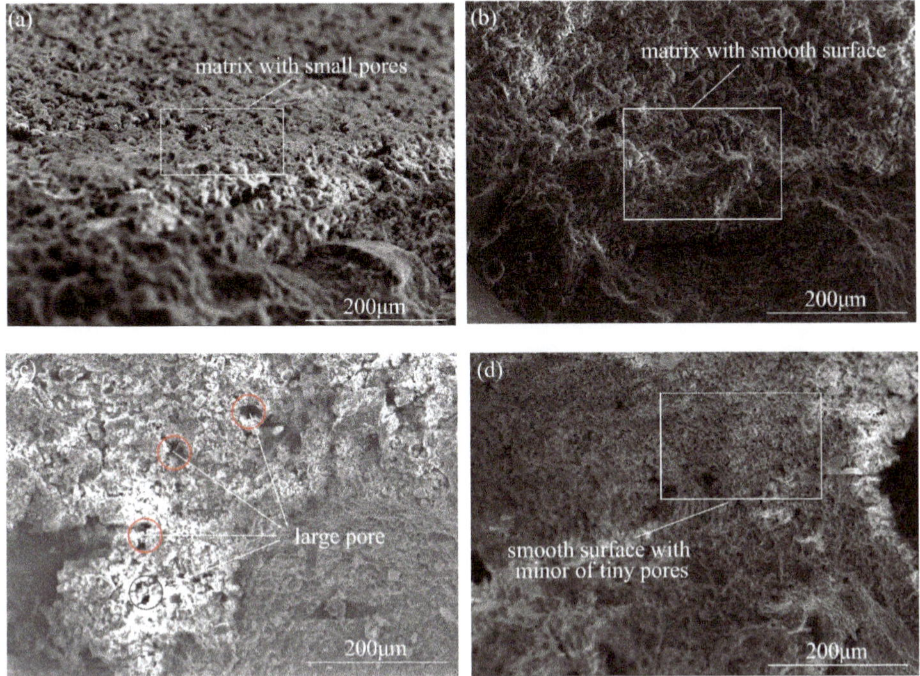

Figure 6. Microstructures of the matrix from SFRC and ZPP-SFRC before and after corrosion: (**a**,**b**) before corrosion, (**c**,**d**) after corrosion.

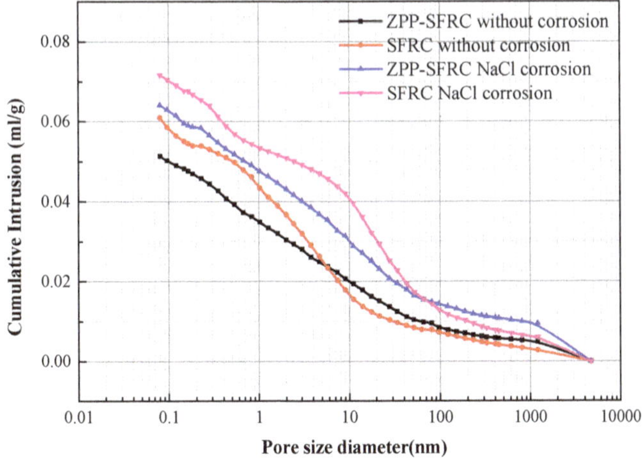

Figure 7. Mercury intrusion of ZPP-SFRC and SFRC.

The three-dimensional and plane morphology of the steel fiber that pulled out from the concrete was observed by Atomic Force Microscope (AFM, Figure 8). The surface of plain steel fiber is relatively flat, with little fluctuation, as compared to zinc phosphate modified steel fiber. However, the overall thickness of zinc phosphate modified steel fiber is relatively higher, the surface undulation has an obviously cone-shaped structure, and the particles are closely arranged due to the existence of the zinc phosphate coating.

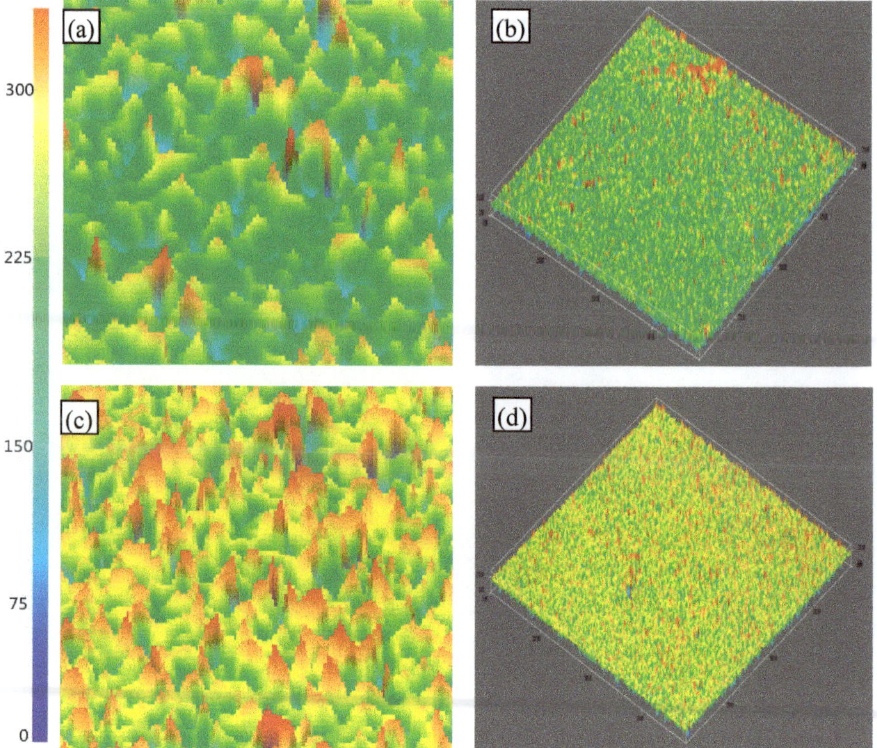

Figure 8. Morphology of plain steel fiber and zinc phosphate modified steel fiber: (**a**) 2D morphology of plain steel fiber, (**b**) 3D morphology of plain steel fiber, (**c**) 2D morphology of phosphatized steel fiber, and (**d**) 3D morphology of zinc phosphate modified steel fiber.

Figure 9 shows the micrographs of plain steel fiber and zinc phosphate modified steel fiber taken out from the crushed concrete before and after corrosion. The internal pH of concrete was around 12.5–13.5 in the alkaline environment, in which steel fiber forms a passive film on its surface to protect itself. The micrograph of the steel fiber from the concrete without corrosion (Figure 9a) indicated that the surface of the steel fiber was relatively clean and there was almost no cement hydration product attached to it. The content of Ca was only 5.53% and that of Fe was 84.14%, indicating that the bond strength between steel fiber and concrete matrix was weak. There was almost no cohesiveness between hydration products and steel fiber, where the bond strength of the interface region mostly came from the cohesive force between the fiber and the matrix. The micrograph of the surface of the plain steel fiber (Figure 9b) was uneven due to NaCl corrosion, and there was a small amount of cement hydration product attached to the fiber due to the rougher surface. The content of Ca increased to 9.91%, Si increased to 4.41%, and Fe decreased to 76.53%.

It can be seen from the micrograph of the zinc phosphate modified steel fiber before and after the corrosion (Figure 9c,d) that, although there were microcracks on the surface of the phosphatized steel fiber after the corrosion, the overall surface morphology did not change significantly. The surface of steel fiber tended to be rough due to the existence of zinc phosphate coating. The surface of zinc phosphate modified steel fiber mainly contains O, Si, P, Ca, Fe, Zn elements (Table 5), indicating that the phosphating treatment of steel fiber has a great influence on the composition of elements in the interface area. In the samples without NaCl corrosion, the content of Ca element in zinc phosphate modified steel fiber increased by 385.35%, and the content of Fe element decreased by 80.21%, compared to the plain steel fiber. The same phenomenon can be found in the samples with NaCl corrosion: the content of Ca element in zinc phosphate modified steel fiber increased by 261.76%, and the content of Fe element decreased by 87.55%. The results show that the interface layer between phosphatized steel fiber and concrete matrix does not only contain conventional hydration products such as C–S–H and Ca(OH)$_2$. Studies [14] have shown that phosphatized steel fiber can contribute to the formation of hydroxyapatite (Ca$_5$(PO$_4$)$_3$(OH)) and calcium phosphate (CaHPO$_4$·2H$_2$O) in concrete. These products can fill the pores of the interface layer between the steel fiber and concrete matrix, which is also an important factor to improve the interface adhesion of the steel fiber and concrete matrix. In the mixing process of concrete, the passivation reaction does not occur on the steel fibers, but a more complex reaction takes place with water and concrete particles due to the presence of the zinc phosphate coating. This reduces the anisotropy between different materials, reduces the thickness of the interface layer, and improves the strength of the weakest point of the interface layer. When bearing the external load, the cracks can be prevented to some extent due to the increase in the bond strength of the interface. When cracks occur, the surface roughness of the phosphatized steel fiber can increase the work needed to pull out the fiber, achieving a toughening effect.

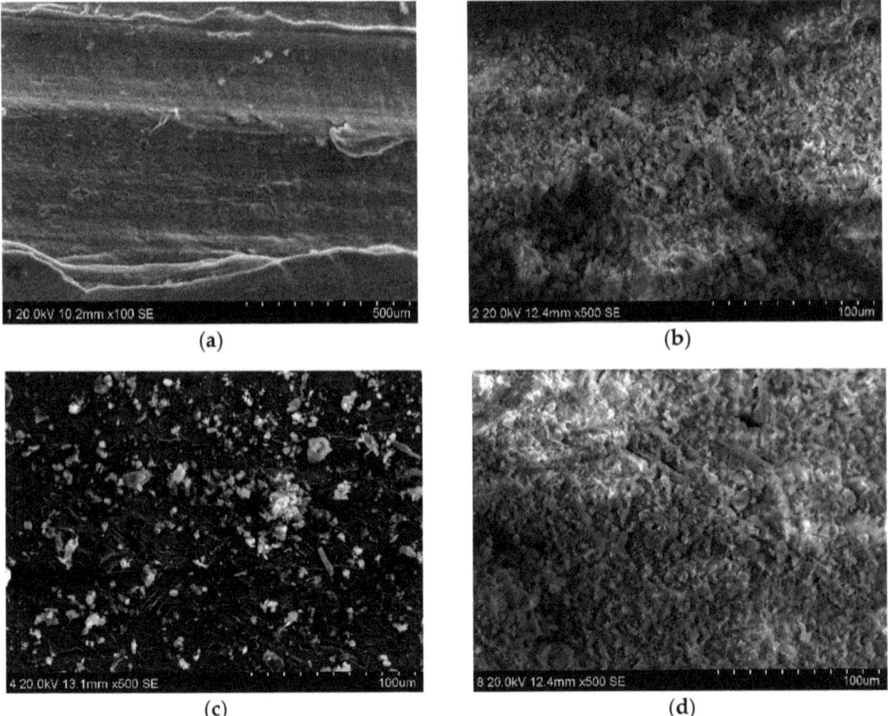

Figure 9. Micrograph of steel fiber pulled out from the concrete: (**a**) before corrosion, plain steel fiber; (**b**) before corrosion, zinc phosphate modified steel fiber; (**c**) after corrosion, plain steel fiber; (**d**) after corrosion, zinc phosphate modified steel fiber.

Table 5. EDS of two kinds of steel fiber pulled out from the concrete before and after corrosion.

Types of Fiber	Environment	Elements						
		O	Al	Si	P	Ca	Fe	Zn
Plain	a	5.58	1.28	1.84	-	5.53	84.14	-
	b	8.54	0.61	4.41	-	9.91	76.53	-
Zinc phosphate treatment	a	22.65	2.43	7.5	5.86	26.84	16.65	17.07
	b	23.83	1.87	6.86	9.76	35.85	9.53	12.3

a Without corrosion, b NaCl solution.

4. Conclusions

This paper provides a method to improve the chloride ion corrosion resistance of steel fiber reinforced concrete. Interfacial bond strength, micro hardness, and micro-morphology properties of the concrete were respectively analyzed. The main findings are as follows:

(1) Phosphating treatment on steel fiber can increase the maximum load during the steel fiber pull-out of concrete in the presence of chloride ions. Furthermore, the bond strength between the steel fiber and concrete matrix is improved. The average bond strength of steel fiber reinforced concrete modified by zinc phosphate increased by 15.4% under the attack of chloride ion.

(2) Phosphating treatment decreases the thickness of the interface layer between steel fiber and concrete by 20 μm, as well as increasing its micro-hardness by around 20 HV, in both corroded and uncorroded environments. With regard to the plain steel fiber reinforced concrete, the presence of a corroded environment increases the thickness of the interface layer by 20 μm and decreases the micro-hardness by 15 HV, compared to that under an uncorroded environment. However, NaCl has little effect on the thickness and micro-hardness of the interface between steel fiber and concrete matrix in the steel fiber reinforced concrete modified by zinc phosphate.

(3) The overall surface morphology did not change much for the zinc phosphate modified steel fiber before and after corrosion, regardless of whether there were minor microcracks on the surface of the phosphatized steel fiber after corrosion.

(4) The steel fiber reinforced concrete modified by zinc phosphate shows good resistance to chloride ion corrosion.

Author Contributions: Writing—original draft preparation and methodology, X.Z.; supervision, R.L.; data curation, W.Q.; writing—review and editing, Y.Y. All authors have read and agreed to the published version of the manuscript.

Funding: This research was funded by the National Natural Science Foundation of China (No. 51902212), Innovation Team of Liaoning Higher Education Institutions (No. LT2019012), Young Top Talents of Liaoning Province (No. XLYC 1807096), as well as the Shenyang Young and Middle-Aged Science and Technology Innovation Talent Support Program (No. RC190374).

Conflicts of Interest: The authors declare that they have no known competing financial interests or personal relationships that could have appeared to influence the work reported in this paper.

References

1. Eik, M. *Orientation of Short Steel Fibres in Concrete: Measuring and Modelling*; Aalto University Publishing: Espoo, Finland, 2014.
2. Cao, J. *Dynamic Constitutive Model and Finite Element Method of Steel Fiber Reinforced Concrete*; Southwest Jiaotong University: Chengdu, China, 2011.
3. Hou, L.; Ye, Z.; Zhou, B.; Shen, C.; Aslani, F.; Chen, D. Bond behavior of reinforcement embedded in steel fiber reinforced concrete under chloride attack. *Struct. Concr.* **2019**. [CrossRef]
4. Cao, H.; Zhang, X.L.; Guo, X.H.; Feng, J.J. Experimental Study on Anti-Penetration Performance of Steel Fiber Reinforced Polymer-Modified Concrete. *Adv. Mater. Res.* **2014**, *1030*, 1100–1103. [CrossRef]
5. Tran, N.T.; Pyo, S.; Kim, D.J. Corrosion resistance of strain-hardening steel-fiber-reinforced cementitious composites. *Cem. Concr. Compos.* **2015**. [CrossRef]

6. Granju, J.L.; Balouch, S.U. Corrosion of steel fibre reinforced concrete from the cracks. *Cem. Concr. Res.* **2005**, *35*, 572–577. [CrossRef]

7. Marcos-Meson, V.; Michel, A.; Solgaard, A.; Fischer, G.; Edvardsen, C.; Skovhus, T.L. Corrosion resistance of steel fibre reinforced concrete—A literature review. *Cem. Concr. Res.* **2017**, *103*, 1–20. [CrossRef]

8. Zhou, M.; Gao, H.J. Study on mix proportion design and construction method of shotcrete. *Ind. Archit.* **2010**, *40*, 80–84.

9. Larisa, U.; Solbon, L.; Sergei, B. Fiber-reinforced Concrete with Mineral Fibers and Nanosilica. *Procedia Eng.* **2017**, *195*, 147–154. [CrossRef]

10. Domski, J.; Katzer, J.; Zakrzewski, M.; Ponikiewski, T. Comparison of the mechanical characteristics of engineered and waste steel fiber used as reinforcement for concrete. *J. Clean. Prod.* **2017**, *158*, 18–28. [CrossRef]

11. Raczkiewicz, W. The Effect of Micro-reinforcement Steel Fibers Addition on the Size of the Shrinkage of Concrete and Corrosion Process of the Main Reinforcement Bars. *Procedia Eng.* **2017**, *195*, 155–162. [CrossRef]

12. Yoo, D.; Yoon, Y.S. Influence of steel fibers and fiber-reinforced polymers on the impact resistance of one-way concrete slabs. *J. Compos. Mater.* **2014**, *48*, 695–706. [CrossRef]

13. Shi, C.; Shao, Y.; Wang, Y.; Meng, G.; Liu, B. Influence of submicron-sheet zinc phosphate synthesised by sol–gel method on anticorrosion of epoxy coating. *Prog. Org. Coat.* **2018**, *117*, 102–117. [CrossRef]

14. Sugama, T.; Carciello, N. Sodium phosphate-derived calcium phosphate cements. *Cem. Concr. Res.* **1995**, *25*, 91–101. [CrossRef]

15. Abosrra, L.; Ashour, A.F.; Youseffi, M. Corrosion of steel reinforcement in concrete of different compressive strengths. *Constr. Build. Mater.* **2011**, *25*, 3915–3925. [CrossRef]

16. Neville, A. Chloride attack of reinforced concrete: An overview. *Mater. Struct.* **1995**, *28*, 63–70. [CrossRef]

17. Montes, P.; Bremner, T.W.; Lister, D.H. Influence of calcium nitrite inhibitor and crack width on corrosion of steel in high performance concrete subjected to a simulated marine environment. *Cem. Concr. Compos.* **2004**, *26*, 243–253. [CrossRef]

18. Tang, L.; Nilsson, L. A discussion of the paper "calculation of chloride diffusivity in concrete from migration experiments, in non-steady-state conditions" by C. Andrade, D. Cervigón, A. Recuero and O. Río. *Cem. Concr. Res.* **1995**, *25*, 1133–1137. [CrossRef]

19. Atis, C.D.; Karahan, O. Properties of steel fiber reinforced fly ash concrete. *Constr. Build. Mater.* **2009**, *23*, 392–399. [CrossRef]

20. Pavia, S.; Condren, E. Study of the Durability of OPC versus GGBS Concrete on Exposure to Silage Effluent. *J. Mater. Civ. Eng.* **2008**, *20*, 313–320. [CrossRef]

21. Oner, A.; Akyuz, S. An experimental study on optimum usage of GGBS for the compressive strength of concrete. *Cem. Concr. Compos.* **2007**, *29*, 505–514. [CrossRef]

22. Narayanan, T.S.N.S. Surface Pretreatment by Phosphate Conversion Coatings—A Review. *Rev. Adv. Mater. Sci.* **2005**, *9*, 130–177.

23. Herschke, L.; Lieberwirth, I.; Wegner, G. Zinc phosphate as versatile material for potential biomedical applications Part II. *J. Mater. Sci. Mater. Med.* **2006**, *17*, 95–104. [CrossRef]

24. Alswaidani, A.M. Inhibition Effect of Natural Pozzolan and Zinc Phosphate Baths on Reinforcing Steel Corrosion. *Int. J. Corros.* **2018**, *2018*, 1–18. [CrossRef]

25. Choe, H.B.; Nishio, Y.; Kanematsu, M. An Investigation on the Usability of Acceleration Test by Impressed Anodic Current for Evaluating Corrosion Behavior of Hot-Dip Galvanized Rebar in Concrete. *Materials* **2019**, *12*, 3566. [CrossRef] [PubMed]

26. Frazão, C.M.V.; Barros, J.A.O.; Bogas, J.A. Durability of Recycled Steel Fiber Reinforced Concrete in Chloride Environment. *Fibers* **2019**, *7*, 111. [CrossRef]

27. Poorqasemi, E.; Abootalebi, O.; Peikari, M.; Haqdar, F. Investigating accuracy of the Tafel extrapolation method in HCl solutions. *Corros. Sci.* **2009**, *51*, 1043–1054. [CrossRef]

 *materials* 

Article

An 18-Month Analysis of Bond Strength of Hot-Dip Galvanized Reinforcing Steel B500SP and S235JR+AR to Chloride Contaminated Concrete

Mariusz Jaśniok *, Jacek Kołodziej and Krzysztof Gromysz

Faculty of Civil Engineering, Silesian University of Technology, 5 Akademicka, 44-100 Gliwice, Poland; jacek.kolodziej@polsl.pl (J.K.); krzysztof.gromysz@polsl.pl (K.G.)
* Correspondence: mariusz.jasniok@polsl.pl

Abstract: This article describes the comparative analysis of tests on bond strength of hot-dip galvanized and black steel to concrete with and without chlorides. The bond effect was evaluated with six research methods: strength, electrochemical (measurements of potential, EIS and LPR), optical, and 3D scanning. The tests were conducted within a long period of 18 months on 48 test elements reinforced with smooth rebars ϕ8 mm from steel grade S235JR+AR and ribbed rebars ϕ8 mm and ϕ16 mm from steel grade B500SP. The main strength tests on the reinforcement bond to concrete were used to compare forces pulling out galvanized and black steel rebars from concrete. This comparative analysis was performed after 28, 180, and 540 days from the preparation of the elements. The electrochemical tests were performed to evaluate corrosion of steel rebars in concrete, particularly in chloride contaminated concrete. The behaviour of concrete elements while pulling out the rebar was observed using the system of digital cameras during the optical tests. As regards 3D scanning of ribbed rebars ϕ8 mm and ϕ16 mm, this method allowed the detailed identification of their complex geometry in terms of determining the polarization area to evaluate the corrosion rate of reinforcement in concrete. The test results indicated that the presence of zinc coating on rebars had an impact on the parameters of anchorage. In the case of ribbed rebars of 16 mm in diameter, the maximum values of adhesive stress and bond stiffness were reduced over time when compared to black steel rebars. Moreover, it was noticed that the stiffness of rebar anchorage in chloride contaminated concrete was considerably higher than in concrete without chlorides.

Keywords: concrete; reinforcing steel; bond strength; zinc coating; HDG; corrosion; chlorides; pullout; EIS; LPR; optical measurements; 3D scanning

Citation: Jaśniok, M.; Kołodziej, J.; Gromysz, K. An 18-Month Analysis of Bond Strength of Hot-Dip Galvanized Reinforcing Steel B500SP and S235JR+AR to Chloride Contaminated Concrete. *Materials* **2021**, *14*, 747. https://doi.org/10.3390/ma14040747

Academic Editor: Luigi Coppola
Received: 31 December 2020
Accepted: 29 January 2021
Published: 5 February 2021

Publisher's Note: MDPI stays neutral with regard to jurisdictional claims in published maps and institutional affiliations.

1. Introduction

The bond strength of reinforcing steel to concrete is a key element that determines the desired behaviour of a reinforced concrete composite. The effects observed at the interface between steel and concrete have been generally thoroughly investigated and clearly described [1–5]. However, introducing a galvanized concrete element to the reinforcement causes some disturbances at the metal-concrete interface. The standard explanation with reference to the mechanical aspects should also include electrochemical issues [6–8]. Three fundamental processes are generally involved in transmitting stress from the rebar to concrete. They are: chemical adhesion, friction between steel and concrete, and mechanical interaction occurring between the reinforcement and the cover [1,9]. Two first processes play an important role for bond strength between the smooth rebars. On the other hand, the mechanical interaction has a fundamental significance for the bond strength with respect to ribbed steel [10]. Thus, the bond stress for smooth steel rebars was only from one-third to one-half of stresses determined for the ribbed rebars [11].

Description of the contact between the rebar and concrete from the electrochemical point of view always requires with analysis of the bond between the metal and liquid filling

81

concrete pores [12–15]. In the case of black steel, the highly alkaline pH of the pore solution (pH > 12.5) leads to the formation of an undesirable passive layer [16–18], whose presence is neglected in the analyses of the adhesion effect. Considering the hot-dip galvanized steel, the high pH of the pore solution creates a direct risk of partial dissolution of zinc coating, particularly for highly alkaline cements [19–21]. The products from zinc corrosion penetrate into concrete pores that have direct contact with the rebar. Additionally, the presence of chlorides in concrete, which is regarded as a corrosive factor, accelerates the destruction of zinc coating [22]. However, the limit level of chlorides is more favourable to galvanized steel than the black one [23–25]. It is also worth mentioning that the volume of products obtained from chlorine induced corrosion of zinc is lower compared to black steel [26]. It leads to a situation in which lower tensile stresses are noticed at the same level of chloride concentrations in the immediate vicinity of the rebar. Consequently, the formation of micro-cracks and the gradual breaking of concrete cover are slowed down [27–29]. All of the above electrochemical aspects of the contact between concrete and galvanized or black steel rebars have a considerable effect on the bond, particularly in the presence of chlorides [12,30].

The failure of adhesion related to friction and chemical adhesion is observed during the pull-out of the rebar from concrete under certain loading [31]. The continuity of deformations of concrete and reinforcing steel is interrupted. Thus, the so called original adhesion disappears and the working phase 1 of the anchored reinforcement is completed [9]. The value of movements in this phase was ca. 1 mm [32]. At the same time, the secondary bond is observed for the ribbed rebars. It is based on transmitting the force from the rebar through its ribs to concrete. The described effect happens because of the two following reasons: necking of the reinforcement caused by increasing stresses and exceeded local pressure of ribs to concrete. Going beyond the maximum pressure values results in the formation of diagonal cracks running from the ribs. Consequently, one small crack is running from each rib of the anchored rebar under correspondingly high stresses and at reduced deformations of the reinforcement cover [32,33]. This level of loading is known as the working phase 2. A further increase in stresses in the rebar leads to the working phase 3 of the anchored rebar. In this phase concrete is spalling under the ribbed area and slipping of the reinforcement is increasing with reference to the surrounding concrete [34]. Therefore, further loading usually causes breaking of concrete cover. So, the broken cover does not function as the closed concrete ring, but as the concrete cantilever. A certain shift of the rebar in the concrete duct corresponds to each working phase [35]. This movement is related to adhesive stress f_b generated on the rebar surface [1]. The value f_b is the highest at the end of the loading phase 3, that is, at the moment of cover cracking.

The values of individual components of stress f_b and $f_\perp$ depend on the angle at which the reaction under the ribs is induced. The value of this angle is variable and within the range of 45–80° [33], depending on the force value exerted on the rebar [36] and the anchorage section. The angle value is not affected by ribbing of reinforcing steel provided that the rib grade exceeds 45° [37]. It can be explained by the deposition of spalling concrete under the ribs which results in the formation of the area inclined at ca. 45° with reference to the axis of the reinforcement [31,34]. It should be emphasized that hot-dip galvanization usually generates a thicker layer of zinc in the ribbed area, which can slightly affect the inclination of the ribs [38]. Then, a small change in the stress values f_b can be noticed compared to the rebar without zinc coating. The component $f_\perp$ can be considered as the hydrostatic pressure acting on the concrete cover from the inside. The analyses of effects observed in the concrete cover are described in the papers, inter alia [31,39].

Many tests have been recently conducted on the effect of corrosion in reinforcement on the maximum values of adhesive stresses. The tests performed on anchored rebars of 12 mm and 16 mm in diameter indicate that bond was increasing at the beginning of the corrosion process [40]. Similar tests on anchored rebars subjected to corrosion for 17 months, which led to the development of cracks of 0.7 mm in width, did not reveal any reduced resistance of anchorage [41]. However, the tests [3] demonstrated the presence of internal cracks of 0.03–0.04 mm in width which were caused by corrosion. They were found to have a significant impact (from 44% to 54%) on reducing the resistance of bond stresses. The investigated anchorages of rebars of 12 mm in diameter indicate that the corrosion at the weight loss of the rebar at the level of 0.5–0.6% caused an increase in bond by 50–60%. On the other hand, the corrosion at the weight loss of the rebar at the level of ca. 1.5% reduced the adhesive resistance by ca. 40% [42]. The corrosion of stirrups was found to have an important effect on the reduced values of adhesive stress. Consequently, greater and adverse effects on the concrete cover were observed [43,44].

According to the comparison in the paper [6], the analyses of bond strength of galvanized steel to concrete, which have been published in recent 100 years, usually focus on short-term tests, e.g., after 28 days of concrete hardening and cover a relatively small number of specimens. However, the problem of the effect of increasing strength of concrete, and the long-term impact of chlorides on the bond of galvanized reinforcement to concrete, have still not been fully examined. Hardly any published papers on testing bond strength of hot-dip galvanized reinforcing steel grade B500SP and S235JR+AR to concrete can be found. In particular, no long-term tests have been so far conducted to analyse how the development of corrosion of galvanized and black steel rebars affected the anchorage stiffness in concrete. Therefore, the tests were also performed to broaden this knowledge.

2. Materials

These tests included the test elements as shown in Figure 1, which were composed of steel rebars *1* embedded in rectangular concrete blocks *2* with dimensions of 150 mm × 150 mm × 130 mm. To reduce the reinforcement contact with concrete to 70 mm in length, a polyurethane foam sleeve *3* with mineral filler was inserted at the section of 60 mm to segmentally separate steel from concrete. In total, 48 test elements were prepared from the reference concrete [45]. Concrete mix with water-cement ratio w/c = 0.45 contained cement CEM I 42.5-SR3/NA, sand, and aggregates with grains of 2–8 mm in diameter. The 1 m^3 of the concrete mix was made up of 489 kg of cement and 1669 kg of aggregates. For 12 test elements, 3% CaCl$_2$ by cement weight was added to the concrete mix to initiate corrosion. The compressive strength of concrete was determined by testing 18 cylindrical specimens with a diameter of 150 mm and a height of 300 mm. After 28 days of concrete hardening, the average strength f_{cm} was determined from 6 cylindrical specimens and amounted to 38.22 MPa, after 180 days—47.70 MPa, and after 540 days—50.16 MPa. The detailed results are shown in Table A1, Appendix A.

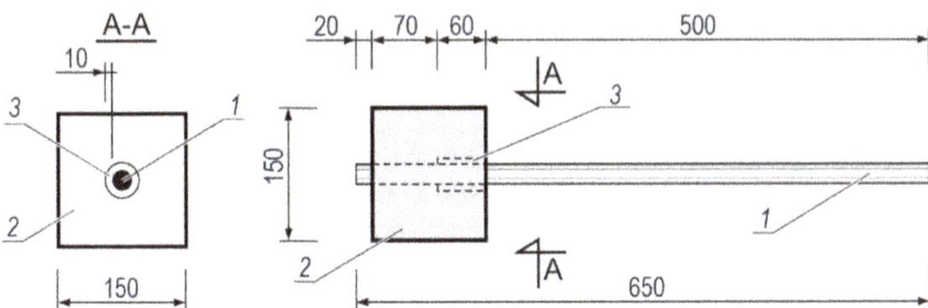

Figure 1. The test element used to test bond of galvanized and black steel to concrete with or without chlorides: *1*—steel rebar, *2*—concrete specimen, *3*—a 10 mm thick sleeve made of polyurethane foam with mineral filler (dimensions in millimetres).

The test elements were reinforced with ribbed and smooth rebars made of steel grade B500SP and S235JR+AR, respectively. Both steel grades are classified as low-carbon steel whose chemical composition is shown in Table 1.

Table 1. Maximum content of elements in the chemical composition of two tested steel grades B500SP and S235JR+AR.

Steel Grade	C	Mn	Si	P	S	Cu	N
B500SP	0.24%	1.65%	0.60%	0.06%	0.06%	0.85%	0.01%
S235JR+AR	0.17%	1.40%	–	0.045%	0.045%	–	0.009%

Figure 2 presents the "stress f–strain ε" graphs for experimentally determined ductility of all types of rebars subjected to the tests. The tests were performed on 18 specimens of reinforcement, six of each steel grade and diameter. Each group of six reinforcement specimens contained two rebars that were a part of the test elements-after their pull-out from concrete. The typical yield strength f_{yk} for all test specimens of steel was within the range of 400–600 MPa, which is recommended by Eurocode [46]. The ductility index of this type of steel can be generally attributed to class C [46] that implies the typical strain ε_{uk} $\geq 7.5\%$ corresponding to the maximum force. Numerical values for strength and tensile behaviour of rebars subjected to the tests are listed in Table A2, Appendix A. They were determined using the analysis of graphs shown in Figure 2.

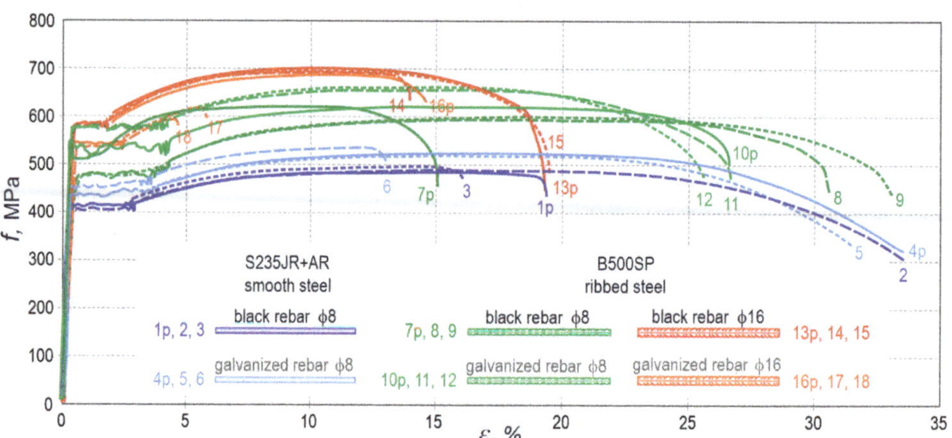

Figure 2. Stress f–strain ε relationship for six types of rebars tested: smooth and ribbed, galvanized, and black steel, with a diameter φ8 mm and φ16 mm, composed of two steel grades; solid lines—the rebars tested after pulling out the specimens from concrete (1p, 4p, 7p, 10p, 13p, 16p), dashed lines—the rebars that were not used in the test elements (2, 3, 5, 6, 8, 9, 11, 12, 14, 15, 17, 18).

The reinforcement used in the specimens was composed of 12 rebars with a diameter φ8 made of smooth steel S235JR+AR, 12 rebars with a diameter of φ8 and 24 rebars with a diameter of φ16—made of ribbed steel B500SP. Half of 48 rebars were galvanized in molten zinc at a temperature of ca. 450 °C prior to setting them in concrete. Zinc bath for 120 s produced a zinc coating of ca. 100 μm in thickness. The detailed analysis of the quality of the zinc coating on ribbed rebars of the same batch and made of steel grade B500SP is presented in the paper [38]. It also includes the effects of chloride-induced degradation of the coating.

3. Methods

The tests were conducted in three stages using six research methods: (1) strength tests on rebar bond to concrete, (2) optical tests and (3) 3D scanning. Moreover, three types of electrochemical tests were performed: (4) the measurements of potential, (5) Electrochemical Impedance Spectroscopy (EIS), and (6) Linear Polarization Resistance

(LPR) technique. Differences in the specimens, conducted stages and the applied research methods are schematically presented in Figure 3. The specimens were numbered from 1 to 24. The independent numbering was applied to the test elements, in which black steel rebars were anchored (letter B) and to those with the galvanized rebars (letter G). The test elements with added $CaCl_2$, were marked with a symbol containing Cl.

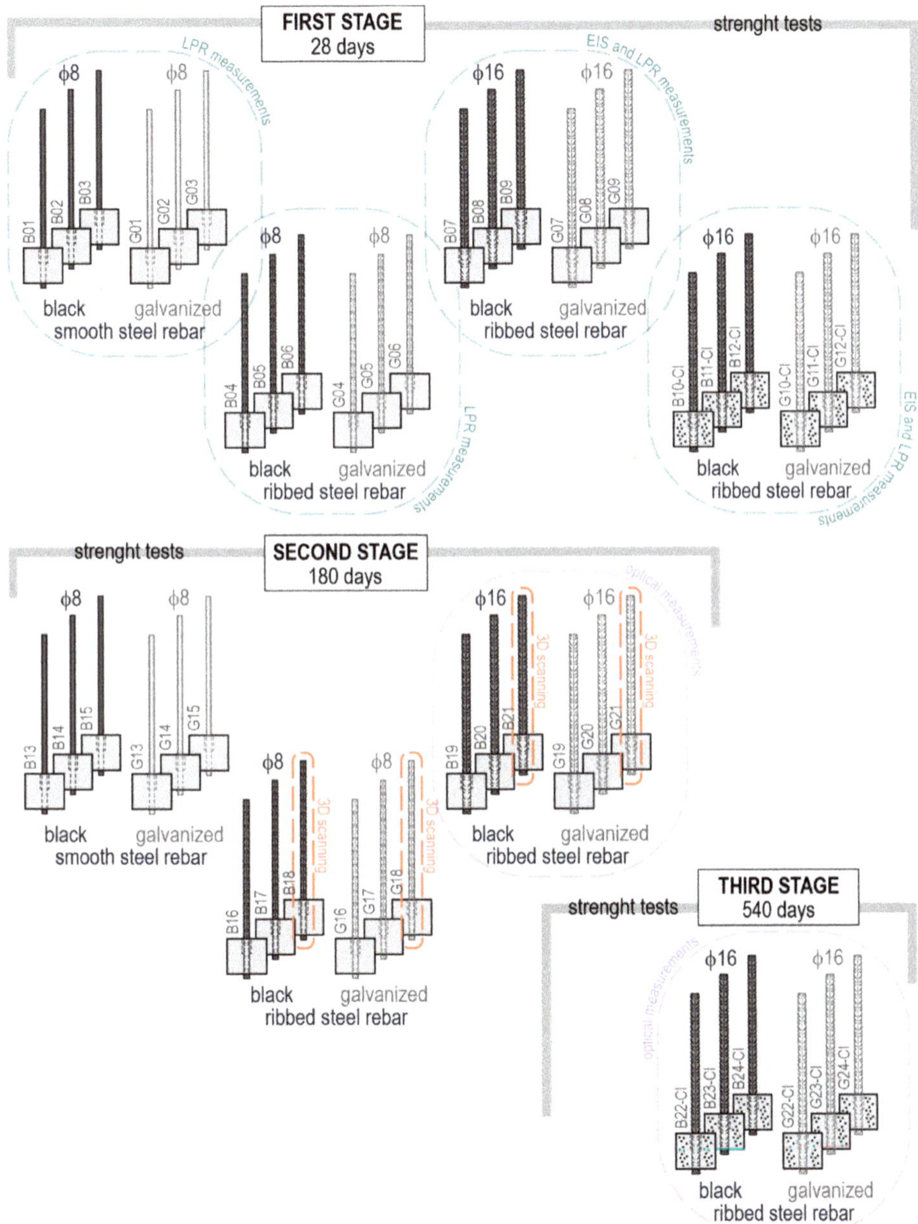

Figure 3. Graphical representation of three test stages of the tests conducted on different types of the test elements with five research methods.

The selected ribbed rebars with a diameter of φ8 mm and φ16 mm were subjected to 3D scanning prior to the preparation of the test elements. The first stage of the tests comprised an analysis of the electrode process at the interface between reinforcing steel and concrete, which was conducted with the electrochemical methods after 3, 10, and 25 days from the preparation of the specimens. On the last day (day 28) of concrete curing in the specimens, a series of 24 measurements of the pull-out force acting on rebars began. The strength was measured for 12 test elements φ8 mm with smooth and ribbed rebars and 12 test elements φ16 mm with ribbed rebars, while six of them were embedded in concrete with chlorides. Half of the 24 rebars were hot-dip galvanized.

The second stage of the tests began 180 days (~6 months) from the preparation of the test elements. This stage included 18 elements: 12 elements φ8 with smooth and ribbed rebars, and six elements φ16 mm with ribbed rebars. All of them were placed in concrete without chlorides. Similarly to the specimens from stage 1, half of the rebars were protected with zinc coating. During the 2nd stage, the main tests on measuring the forces pulling out the rebars from concrete were performed simultaneously with the optical measurements of displacement of points painted on lateral surfaces of the specimens. Due to the limited access to the measuring equipment of the Aramis system, the optical measurements were only performed on rebars φ16 mm at this test stage.

The 3rd stage began 540 days (~18 months) from setting the test elements in concrete. The forces pulling out the rebars from concrete were measured in other six specimens: three of them reinforced with galvanized ribbed rebars and three others reinforced with black steel ribbed rebars, φ16 mm, embedded in concrete with chlorides. The optical measurements using the Aramis camera system were performed in the same way as in stage 2.

3.1. Strength Tests—Measuring the Force Pulling Out the Rebar from Concrete

The test stand for testing the force pulling out the rebars from concrete is shown in Figure 4a, and its diagram is illustrated in Figure 4b. The test element is placed between holders of the stand.

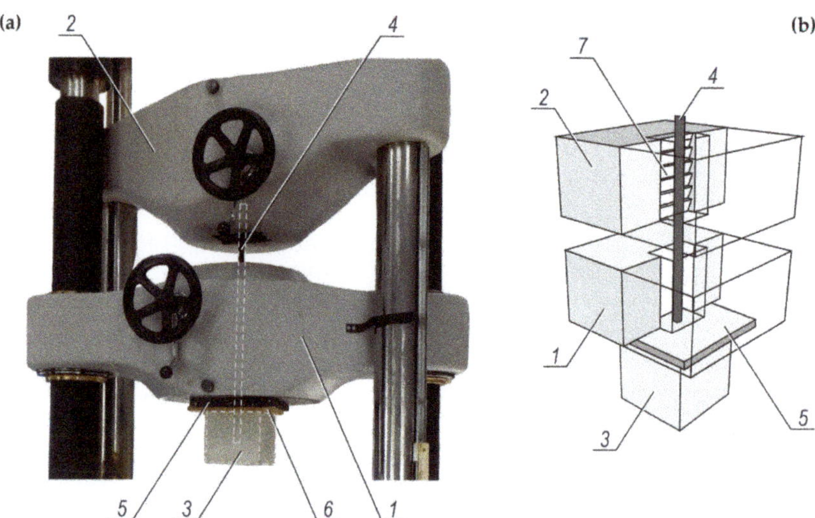

Figure 4. The test stand for testing the force pulling out the rebars from concrete: (**a**) view of the test machine from the Labortech company (Opava, Czech Republic) during the tests, (**b**) a diagram of the test system; *1*—a bottom holder of the test machine, *2*—a top holder of the test machine, *3*—a concrete specimen being a part of the test element, *4*—a rebar placed between holders and pulled out from the concrete specimen, *5*—steel slab transmitting force to the concrete specimen, *6*—equalizing washer made of HDF (high-density fibreboard), *7*—convergent self-clamping grips of the top holder of the test machine.

The tests were performed with the Labortech test machine. The HDF (high-density fibreboard) equalizing washer *6* was placed on the concrete specimen *3*. Then, the steel slab *5*, transmitting uniformly the force to the concrete specimen *3*, was placed on the HDF washer. In this system the steel rebar *4* protruding from the concrete specimen passed through the bottom holder *1* of the test machine, and then was inserted into the top holder *2* with the convergent self-clamping grips *7*.

The forces pulling rebars out of concrete were measured by controlling an increment of the force over time. The force increment was constant during the 1st stage of the tests, that is, after 28 days of concrete hardening in the specimens. It was equal to 1 kN/s for a diameter ϕ16 mm and 0.5 kN/s for ϕ8 mm. Stage 2 of the tests was conducted after 180 days from the day of preparing the test elements. At this stage the force increment was increased over time. However, the measurements were interrupted when the ultimate force reached ca. 70% of the value obtained in the stage 1. The reason for that was the introduction of the Aramis system for optical measurements. The force increment was continued after changing the recording frequency of the cameras.

3.2. Optical Tests—Measuring Displacements of Points with the Aramis Camera System

The Aramis camera system from the company GOM (Braunschweig, Germany) was additionally used to provide more detailed information on the pull-out force (Figure 5a). The system was based on the Digital Image Correlation technique. The optical measurements are non-contact and are insensitive to the behaviour of materials. The system was composed of two 6 Mpx cameras (GOM, Braunschweig, Germany) equipped with 24 mm lenses to measure the test area of 120 mm × 150 mm and 90 mm in depth. The scanned area facilitates the analysis of data collected both from the whole area and the selected points. At a time before the measurements, the surface of concrete specimen *1* was painted white, and then sprayed with graphite paint (Figure 5b) to cover ca. 50% of the surface. Then, a plate *3* with glued permanent measuring points was fixed to the rebar *2*. The measurements were taken for displacements of the bar *3* with respect to the bottom edge of the concrete element, from which the rebar was pulled out.

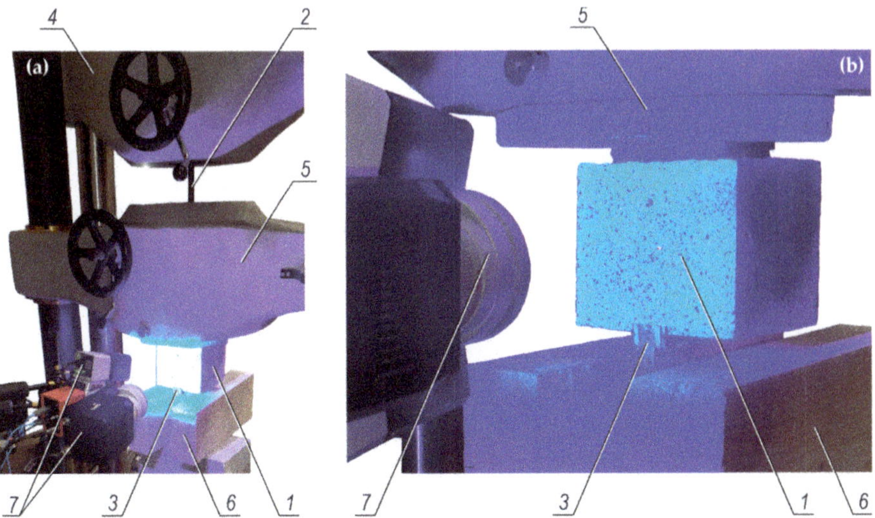

Figure 5. The Aramis system of cameras from the GOM company used on stage 2 and 3 of testing the force pulling out the rebar from concrete: (**a**) view of the whole test stand, (**b**) view during measurements of the concrete specimen with a spot sprayed graphite paint; *1*—concrete specimen, *2*—a steel rebar pulled out from concrete in the specimen, *3*—a bar with glued permanent measuring points, *4*—the top holder of the test machine, *5*—the bottom holder of the test machine, *6*—wood beams to support the concrete specimen after the rebar pullout, *7*—cameras recording the measurement.

3.3. Electrochemical Testing of Corrosion of Concrete Reinforcement

The density of corrosion current was measured for the reinforcement in concrete specimens using a three-electrode arrangement illustrated in Figure 6. Both ends of the steel rebar *1* protruding from the concrete specimen *2* served as the working electrode. The counter electrode *3* made of stainless-steel sheet was placed on a wet felt pad *4*. A third electrode was the reference electrode *5* (Cl⁻/AgCl,Ag) that was stabilised in the ballast guide *6*. All three electrodes were connected to the potentiostat *7* (Gamry Instruments, Warminster, PA, USA)—the Reference 600 model from the Gamry company.

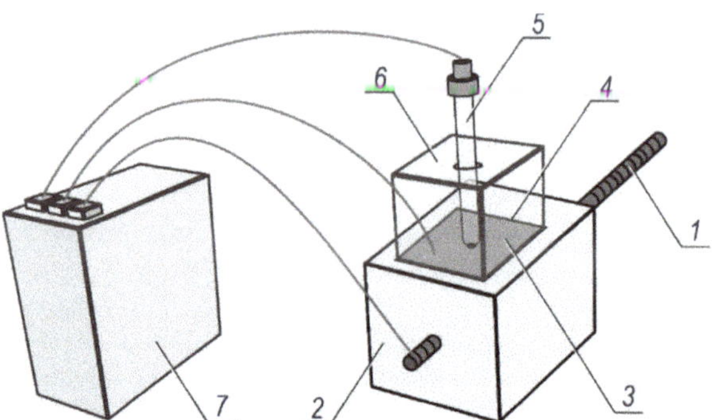

Figure 6. The test set-up used for the polarization tests of corrosion rate in concrete reinforcement: *1*—steel rebar, *2*—concrete specimen, *3*—counter electrode made of stainless steel sheet, *4*—felt pad, *5*—reference electrode Cl⁻/AgCl,Ag, *6*—ballast with a guide for the reference electrode, *7*—potentiostat.

The electrochemical tests on polarization were conducted on a part of the specimens using in the first place the electrochemical impedance spectroscopy (EIS) technique. The samples were stored at ca. 20 ± 2 °C and a relative humidity of $50 \pm 10\%$. Prior to beginning polarization tests on a given day, the top surface of the concrete sample was immersed in tap water to the depth up to 30 ± 1 mm for ca. 15 min to ensure better conductivity of concrete cover in the tested reinforcement during measurements. When the specimens were taken from water, they were connected to the potentiostat. Changes of the gradually stabilising potential were observed using the reference electrode. When changes in the potential were at the level of 0.1 mV/s, the EIS technique was used to perform the tests on the steel reinforcement in concrete. The measurements were taken in the potentiostatic mode at the fixed range of frequencies of 0.01 Hz–100 kHz and a disturbing sinusoidal signal was applied at the potential amplitude of 10 mV over the corrosion potential. After completing the impedance measurements, the changes in the potential of the reinforcing steel in concrete were again observed. The potentiodynamic measurements using the LPR technique were taken when the potential was stabilized at the level of 0.1 mV/s. The reinforcement was polarized at a rate of 1 mV/s within the defined range of potential changes from −150 mV to +50 mV against the corrosion potential.

3.4. 3D Scanning of Ribbed Rebars

The detailed representation of the complex geometry of ribbed rebars embedded in the concrete specimens was mapped in detail using the laser scanner Model Maker MMDx100 from the Nikon Metrology company (Leuven, Belgium). The scanning head MMDx100 based on the ESP3 technology emits a laser beam of 100 mm in width at the declared accuracy of 20 µm. Scanning of the steel ribbed rebars was supported with the 7-axis measurement arm MCAx20 from the Nikon Metrology company (Leuven, Belgium). Its measurement range is 2.0 m, the measurement accuracy of a point is 0.023 mm, and the spatial accuracy is 0.033 mm.

4. Results

4.1. Effects of 3D Scanning of Ribbed Rebars

Figure 7 present 3D images of typical fragments of ribbed rebars with a diameter ϕ8 mm and ϕ16 mm. They were scanned using the scanner (Nikon Metrology, Leuven, Belgium) specified in point 3.4. The obtained point cloud data were processed with the Focus software from the Nikon Metrology company (Leuven, Belgium). The surface of ribbed rebars was calculated using the Alibre Design software from the Datacomp company (Crakow, Poland). The calculated surface area of the ribbed rebar at the interface between reinforcing steel and the concrete specimen over a distance of 70 mm was equal to 20.10 cm^2 for the rebar diameter ϕ8 mm, and 40.21 cm^2 for the rebar ϕ16 mm.

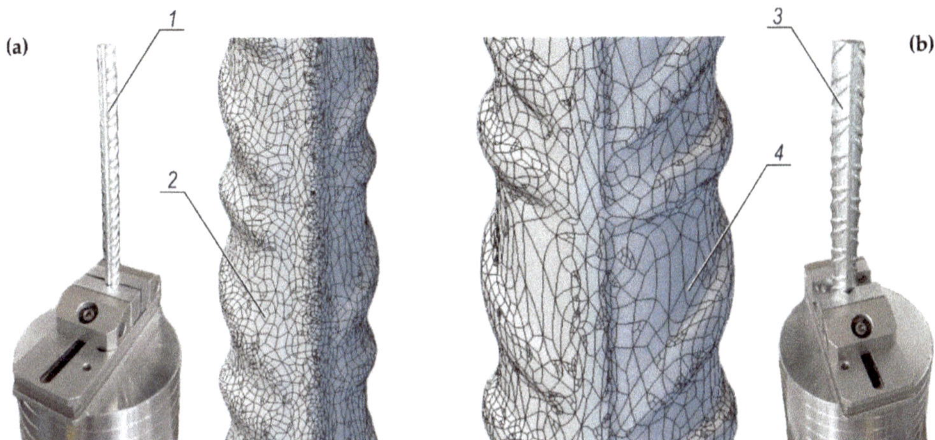

Figure 7. 3D scanning of typical fragments of ribbed rebars made of steel grade B500SP: (**a**) a rebar with a diameter of 8 mm, (**b**) a rebar with a diameter of 16 mm; *1* and *3*—rebars ϕ8 mm and ϕ16 mm stabilised prior to 3D scanning, *2* and *4* —represented geometry of rebars ϕ8 mm and ϕ16 mm based on point cloud data.

The obtained detailed information on the complex geometry of typical fragments of ribbed rebars of both diameters (ϕ8 mm and ϕ16 mm) were used in described, further in this paper, tests to determine the polarized area of the reinforcement part being in contact with concrete from the specimen. This information was regarded as the required one to determine the density of corrosion current during the polarization tests using the EIS and LPR techniques.

4.2. Test Results for Corrosion Potential of Rebars in Concrete

Figure 8 illustrates the results for the average measurements of the corrosion potential E_{corr} for galvanized and black reinforcing steel in 24 specimens subjected to the tests. The tests referred to the first series of the specimens, that is, B01–B12-Cl and G01–G12-Cl. The potential was measured as a preliminary element of the measurement procedure, which was required prior to the polarization tests concerning corrosion rate of reinforcement corrosion conducted with the LPR technique. The potential value E_{corr} was read only when its variability was stabilised at the level of 0.1 mV/s. The potential values expressed as [V] and measured against the reference electrode Cl$^-$/AgCl,Ag are presented in the tabular form in Figure 8a below the bar charts.

The potential was measured each time for three specimens of the same type, and the calculated arithmetic means were inserted into the tables in Figure 8a. The measurements of E_{corr} were repeated three times, that is, on days 3, 10, and 25 after setting the specimens in concrete. Each of three measuring series was marked with a different colour as seen in Figure 8a to facilitate the analysis of obtained results. Moreover, the specimen sketches denoted with the rebar types and specimen symbols are illustrated in Figure 8b to connect the obtained results with particular types of specimens.

The measurement results presented in the form of bar charts demonstrate a clear difference between the potentials E_{corr} for galvanized and black steel. The potential of galvanized rebars was lower by 159–512 m than rebars without the protective coat placed in the same concrete. This can be attributed to zinc's position against iron in the electrochemical series of metals. The potential drop was observed along with concrete hardening. It means that the most negative potentials were generally found for the last measurement series on day 25 from setting the specimens in concrete. The specimens G10-Cl–G12-Cl with galvanized rebars in concrete with chlorides were the exception. A reverse trend was observed for them.

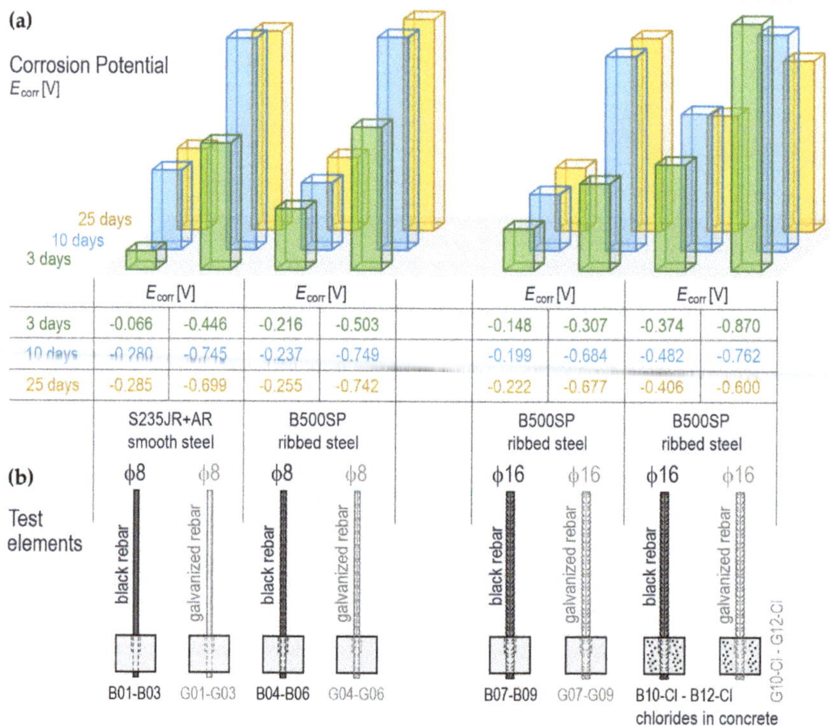

(a)	E_{corr} [V]		E_{corr} [V]			E_{corr} [V]		E_{corr} [V]	
3 days	-0.066	-0.446	-0.216	-0.503		-0.148	-0.307	-0.374	-0.870
10 days	-0.280	-0.745	-0.237	-0.749		-0.199	-0.684	-0.482	-0.762
25 days	-0.285	-0.699	-0.255	-0.742		-0.222	-0.677	-0.406	-0.600
	S235JR+AR smooth steel		B500SP ribbed steel			B500SP ribbed steel		B500SP ribbed steel	
(b) Test elements	ϕ8 black rebar	ϕ8 galvanized rebar	ϕ8 black rebar	ϕ8 galvanized rebar		ϕ16 black rebar	ϕ16 galvanized rebar	ϕ16 black rebar	ϕ16 galvanized rebar
	B01-B03	G01-G03	B04-B06	G04-G06		B07-B09	G07-G09	B10-Cl - B12-Cl	G10-Cl - G12-Cl
								chlorides in concrete	

Figure 8. The results from electrochemical tests on reinforcing steel in concrete on days 3, 10, and 25 from setting the specimens in concrete: (**a**) corrosion potential E_{corr} averaging from three specimens (**b**) the test elements with galvanized and black reinforcing steel (smooth ϕ8 mm and ribbed ϕ8 and ϕ16 mm) in concrete with and without chlorides.

4.3. Impedance Results for Reinforcing Steel in Concrete

Figure 9 illustrates the results for impedance measurements taken for 12 specimens with ribbed rebars ϕ16 mm. Types of the test specimens are schematically presented in Figure 9(a1). The measurements with the EIS technique, similarly to the measurements of potential, were taken in 3 measurement series on days 3, 10, and 25 from the day of preparing the specimens. For better clarity of the presented results, different colours were attributed to the impedance spectra describing four types of the specimens: blue—the specimen with a black steel rebar in concrete without chlorides, green—the specimen with galvanized rebar in concrete without chlorides, red—the specimen with a black steel rebar in concrete with chlorides, violet—the specimen with the galvanized rebar in concrete with chlorides.

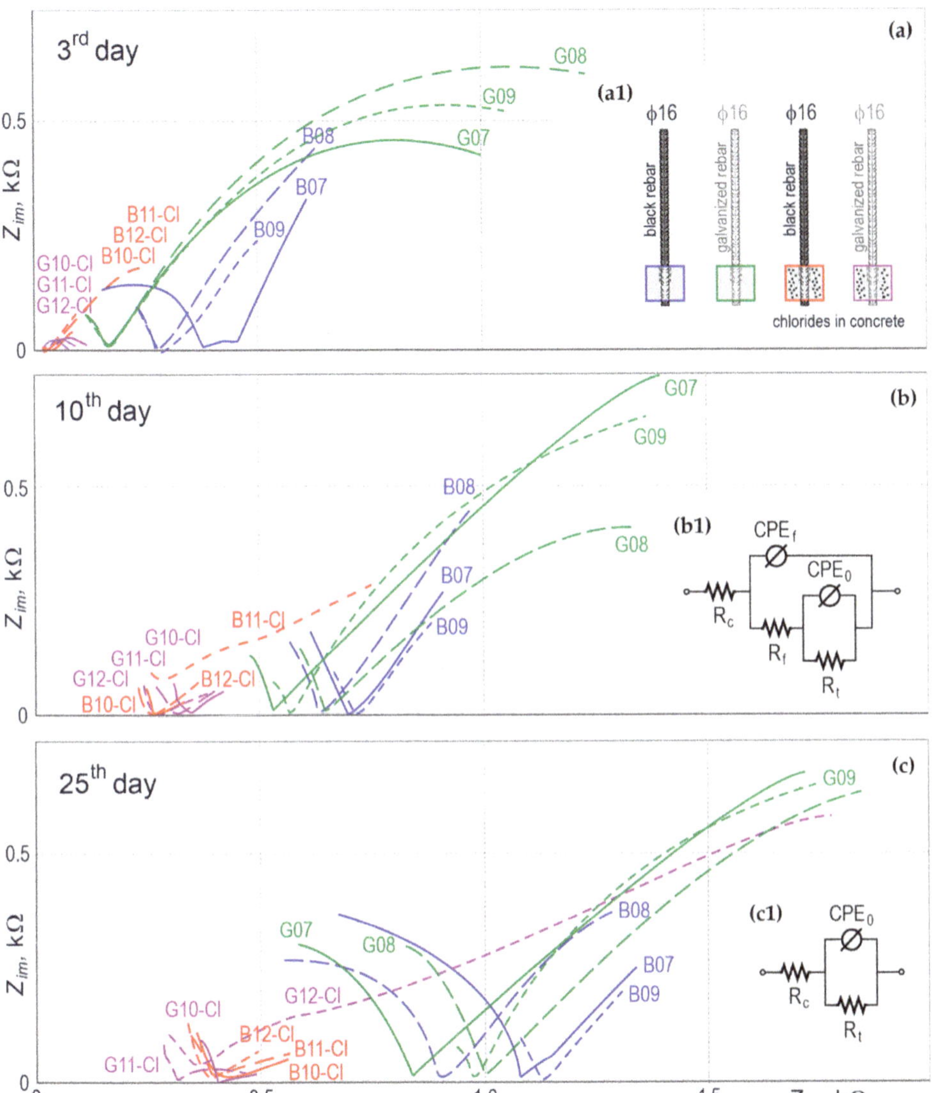

Figure 9. Impedance results on the Nyquist plot for ribbed rebars φ16 mm embedded in the concrete specimens: (**a**) the measurements on day 3 of concrete hardening in the specimens, (**a1**) 4 types of the elements tested with EIS, the colour of concrete paver outline corresponded to the colours of impedance spectra, (**b**) the measurements on day 10 after setting the specimens in concrete, (**b1**) equivalent electrical circuit for analysing the spectra of steel in concrete with chlorides, (**c**) the measurements on day 25 after setting the specimens in concrete, (**c1**) equivalent electrical circuit for analysing the spectra of steel in concrete without chlorides.

The impedance spectra in the complex Z_{re}–Z_{im} plane are illustrated in Figure 9a–c. Their analysis shows that the impedance spectra for galvanized and black steel rebars in concrete with chlorides had considerably lower frequencies compared to the corresponding spectra for concrete without chlorides which indicated lower impedance of the described system. Low values of the impedance modulus in relation to flattened semi-circles (Figure 9a, specimens G10-Cl–G12-Cl) or shallow curves (Figure 9a–c, specimens B10-Cl–B12-Cl and Figure 9b,c, specimens G10-Cl–G12-Cl) within a low-frequency range can indicate the corrosion risk for steel in concrete. The impedance spectra for galvanized and black steel rebars in concrete without chlorides, on the other hand, show a few-fold higher impedance of the system. Two clear fragments can be observed for each spectrum. A curve or a small fragment of the flattened semi-circle was noticed on the left side of the characteristic inflection point of spectra, which was the closest to the horizontal axis of the real impedance. This part of the spectrum with high frequencies described the electrochemical properties of concrete and high-alkali liquid filling its pores. A considerably larger curve or a bigger fragment of the flattened semi-circle was noticed on the right side of the characteristic inflection point of spectra. This part with low frequencies was typical for the passivation of steel in concrete. A flattened shape of the semi-circle could indicate the initiation of gradual depassivation of metal.

The above qualitative analysis of experimentally determined 36 spectra was verified with the quantitative analysis based on adjusting theoretical spectra to corresponding electrical equivalent circuits illustrated in Figure 9(b1,c1). The spectra typical for galvanized and black steel rebars in concrete without chlorides were modelled using the modified Randles circuit (Figure 9(c1)), in which capacity was replaced with the constant phase element (CPE). Considering the spectra of rebars in concrete with chlorides, they were modelled using the circuit presented in Figure 9(b1) because the second time constant was noticed in the range of low frequencies. It should be mentioned that both equivalent electrical circuits were selected regarding only the analysis of low-frequency parts of spectra which were typical for electrochemical properties of metal in concrete. In both circuits (Figure 9(b1,c1)), the resistance R_c specified the impedance of the whole concrete, the resistance R_t specified the charge transfer resistance through the metal-pore solution interface, and the CPE_0 characterized the pseudo-capacity of a double layer on steel. For the scheme illustrated in Figure 9(b1), the resistance R_f and the CPE_f specified the resistance and the pseudo-capacity of the forming layer of corrosion products of iron or zinc, which locally limited the access of oxygen to metal surface.

The detailed results for adjusting individual electrochemical parameters of both electrical equivalent circuits using the Simplex method are presented in Table A4, Appendix B. The charge transfer resistance R_t, which was a key parameter, was used to calculate the density of corrosion current $i_{corr} = B/R_t$ from the Stern-Geary equation. The parameter B was determined from the slope coefficients of linear sections of the polarization curves (anodic b_a and cathodic b_c) obtained during the potentiodynamic tests conducted with the LPR technique described in point 4.4.

4.4. Potentiodynamic Results from Testing Polarization of Reinforcing Steel in Concrete

The potentiodynamic results from the LPR tests for the steel reinforcement in concrete performed after 3, 10, and 25 days from setting the specimens in concrete are shown in Figure 10. Contrary to the impedance tests specified in point 4.3, the direct current polarization measurements were taken for all 24 specimens as part of test stage 1—cf. Figure 3 A total of 72 polarization curves were obtained due to three measurement series taken within a few days at an interval of several days. By determining the polarization resistance R_p for each curve and the slope coefficient of a straight section of the anodic b_a and cathodic curves, the corrosion current density $i_{corr} = B/R$, where the parameter $B = b_a b_c/2.303 (b_a + b_c)$, was calculated from the Stern-Geary equation. The densities of corrosion current averaged each time from three specimens are shown as a bar graph in Figure 10a. Different colours were attributed to individual measurement series for better clarity of the presented results. Additionally, the averaged densities of corrosion current i_{corr} expressed as $[\mu A/cm^2]$, are presented in the tabular form below the bar graph. Figure 10b shows sketches of all types of specimens tested with the LPR technique. They contain information on rebar diameter and steel grade, the presence or absence of the zinc coating and number and letter symbols.

The analysis of corrosion current densities presented that the following values i_{corr} = 0.05–0.13 $\mu A/cm^2$ were obtained from all three measurement cycles for the specimens with black ribbed rebars $\phi 8$ mm and $\phi 16$ mm. This could indicate passivation. For the specimens made of smooth steel rebars $\phi 8$ mm, an increase in i_{corr} = 0.49 $\mu A/cm^2$ was observed within two consecutive measurement cycles apart from the first one. This increase was not entirely explicable; however, it could suggest partial decomposition of the passive layer on metal. The expected increased values i_{corr} = 0.36–0.63 $\mu A/cm^2$, indicating the development of corrosion, were found for unprotected ribbed reinforcing steel $\phi 16$ mm without zinc protection, which was set in concrete with aggressive chloride ions. In the case of the similar specimens but with the galvanized rebar, a dramatic rise in densities of the corrosion current up to i_{corr} = 9.23 $\mu A/cm^2$ was noticed during the 1st cycle of measurements. This value was gradually falling in consecutive measurements to i_{corr} = 2.91 $\mu A/cm^2$ on day 10, finally reaching the value i_{corr} = 0.67 $\mu A/cm^2$ on day 25. It should be emphasized that very close range of corrosion current densities was obtained from the EIS tests, which had been conducted on a part of the specimens prior to the LPR tests. The observed effect of a rapid rise in the density of corrosion current can be attributed to the cumulative impact of chloride ions due to corrosion and high pH of pore solution, which caused partial dilution of the zinc coating at the initial stage of concrete setting. The values of corrosion current density for other specimens with galvanized rebars in concrete without chlorides ranged from 0.05 to 0.42 $\mu A/cm^2$. These values usually did not exceed i_{corr} = 0.20 $\mu A/cm^2$ indicating no signs of corrosion.

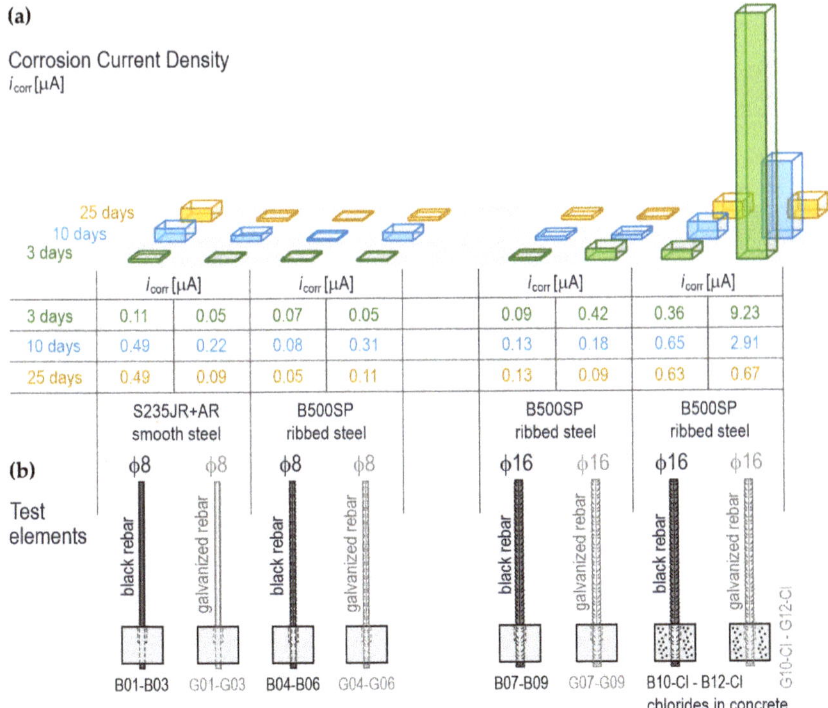

	i_{corr} [µA]		i_{corr} [µA]			i_{corr} [µA]		i_{corr} [µA]	
3 days	0.11	0.05	0.07	0.05		0.09	0.42	0.36	9.23
10 days	0.49	0.22	0.08	0.31		0.13	0.18	0.65	2.91
25 days	0.49	0.09	0.05	0.11		0.13	0.09	0.63	0.67
	B01-B03	G01-G03	B04-B06	G04-G06		B07-B09	G07-G09	B10-Cl - B12-Cl	G10-Cl - G12-Cl

Figure 10. The results from electrochemical tests on reinforcing steel in concrete on days 3, 10, and 25 from setting the specimens in concrete: (**a**) corrosion current density i_{corr} determined with the LPR technique, (**b**) the test elements with galvanized and black reinforcing steel (smooth $\phi 8$ mm and ribbed $\phi 8$ and $\phi 16$ mm) in concrete with and without chlorides.

4.5. Results from Optical Measurements

The optical measurements with the Aramis camera system were additionally taken to provide new and detailed information on the strength tests. Their main objective was to record the displacements of points on the surface of the loaded tests element while pulling out the rebars from concrete causing the specimen failure. These measurements were used to catch the moment of fracture and crack formations in the test element, in which the reinforcement was anchored. Finding the connection between the force N acting on the rebar and its displacements s against the specimen was very important information used to determine the anchorage stiffness of rebar in concrete. Figure 11 presents the reference images of the surface of concrete specimens (containing chlorides) during the pull-out of ribbed rebars ϕ16 mm after 540 days from their preparation. A visible brown scattered grid *3* of displaced points is shown in Figure 11a. This image was captured during the pull-out of the black steel rebar. The pull-out of the galvanized rebar is presented in Figure 11c, where the very condensed brown grid *5* of the displaced points can be noticed. This grid changes into the asymmetrical crack *6* (bursting) of concrete *2* as presented in Figure 11d which illustrates the moment of rapid failure. Similarly, the moment of rapid cracking *4* of the specimen *1* is presented in Figure 11b. The crack symmetry can be related to the scattered, but almost centrally located, brown grid *3* of the displaced points that is shown in Figure 11a.

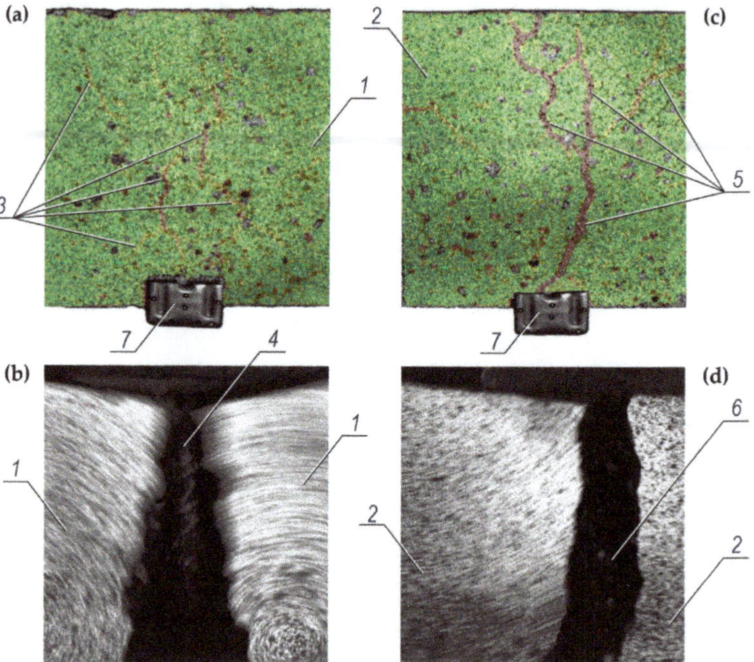

Figure 11. The analysis of the surface of concrete specimens with chlorides during the pull-out of ribbed rebars ϕ16 mm after 540 days from their preparation, which was conducted using the Aramis system: (**a**) surface of the specimen *1* (with a black steel rebar) and hardly visible dispersed brown grid *3* of displaced points sprayed with graphite paint, (**b**) the moment of rapid cracking *4* of concrete in the specimen *1* in the grid zone *3*, (**c**) surface of the specimen *2* (with the galvanized rebar) and very clear condensed brown grid *5* of displaced points, (**d**) the moment of rapid asymmetrical cracking *6* of concrete in the specimen *2* in the grid zone *5*; *7*—plate with permanent measuring points glued to the pulled out rebar.

Figure 12 presents the results from measuring the rebar shift s [mm] from the concrete block caused by an increment of the pull-out force F [kN]. The measured shift was recorded using the Aramis camera system, whose software interacted with the control system for an increase in force of the strength machine. Due to the limited access to the Aramis system, the tests were reduced to galvanized and black steel ribbed rebars ϕ16 mm, embedded in concrete with and without chlorides. The elements without chlorides were tested after 180 days from the day of their preparation, and the elements containing chlorides were tested after 540 days.

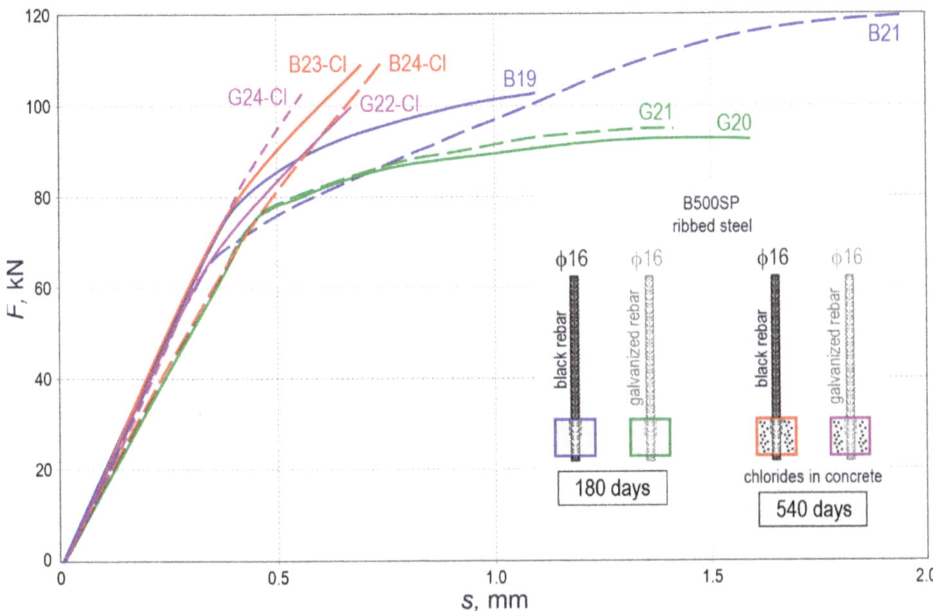

Figure 12. The relationship between the pull-out force F [kN] and the shift s [mm] during the pull-out of the galvanized and black steel rebar from concrete. This relationship, based on the Aramis system, was determined for rebars ϕ16 mm from steel grade B500SP—the tests performed after 180 and 540 days.

The test results shown as curves in Figure 12 indicate the paths $F(s)$ for the load range $F = 0$–60 kN were quasi-linear for each test element. A further increase in load, that is, for $F > 60$ kN, the elements containing chlorides B23-Cl, B24-Cl, G22-Cl, G24-Cl maintained their straight-line relationship $F(s)$. On the other hand, the curves representing the elements (B19, B21, G20, G21) without chlorides in concrete had a clear drop in the slope $F(s)$.

4.6. Results from Measuring the Force Pulling Out Rebars from Concrete

All 48 test elements were damaged. Figure 13 shows the reference images of the damaged test elements after pulling out the ribbed rebars with a diameter of 16 mm. It can be seen that breaking of the concrete specimens was both symmetrical and asymmetrical into two (Figure 13a,b), and even four parts (Figure 13c).

The bar graphs presented in Figure 14 compare values of the forces pulling out the rebars from concrete. The pull-out forces were averaged each time from three specimens. The notations $F_{max,G}$ and $F_{max,B}$ were used to differentiate the pull-out forces acting on the galvanized and black steel rebars, respectively. The values of the pull-out forces F_{max} determined for all 48 specimens are shown in Table A3, Appendix A. Different colours were attributed to determined forces to simplify the analysis of test results shown in Figure 14a: test stage 1—after 28 days from preparing the specimens (green), test stage

2—after 180 days (blue), and tests stage 3—after 540 days (yellow). Additionally, the sketches shown in Figure 14b representing the specimens with black steel reinforcement are in dark grey, and the ones with galvanized rebars are in light grey. To differentiate the results obtained for smooth steel (S235JR+AR) and ribbed steel, (B500SP), the grades and numbers of individual specimens are next to the specimen outlines.

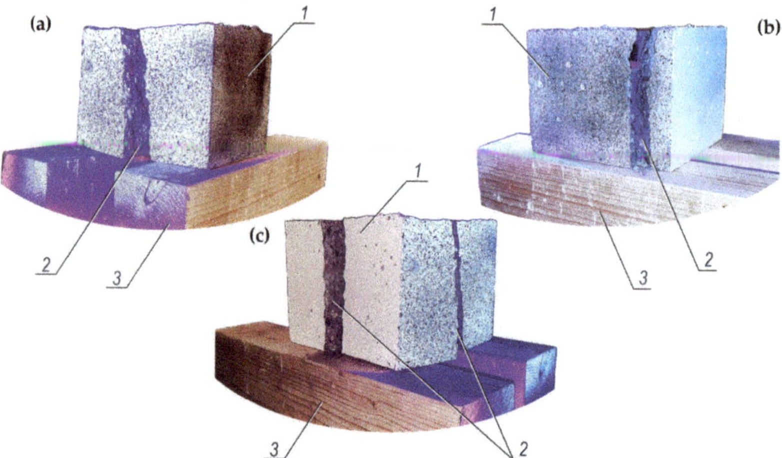

Figure 13. Images of typical failure of concrete specimens after pulling out the ribbed rebars of 16 mm in diameter: (**a**) uniform breaking of the specimen into two parts, (**b**) inclined breaking of the specimen into two parts, (**c**) breaking of the specimen into four parts; *1*—concrete specimen with the pulled-out reinforcement, *2*—cracked location in the specimen, *3*—wooden beams supporting the concrete specimens after pulling out the rebar.

The highest values of the pull-out forces (73.11–110.17 kN) were determined as expected for the ribbed rebars ϕ16 mm, and the lowest values (7.99–19.69 kN) for the smooth rebars ϕ8 mm. Except for the ribbed rebars ϕ8 mm, an increase in the pull-out force was noticed after 180 or 540 days. The greatest rise expressed in percentage was found for the smooth rebars ϕ8 mm. Considering the mean pull-out forces acting on the same specimens, the values obtained for galvanized rebars were generally lower compared to black steel rebars. Other results were obtained for smooth steel ϕ8 mm after 180 from setting the specimens in concrete and ribbed steel ϕ16 mm after 28 days from setting the specimens in concrete. A significant rise in the pull-out forces acting on the ribbed rebars ϕ16 mm after 540 days from preparing the specimens with chloride ions should be noted. These ions were aggressive for the reinforcement, so the specimens were very deteriorated due to corrosion.

Figure 15 presents the images of five ribbed rebars ϕ16 mm after pulling out from the concrete block specimens. Black steel *1* and galvanized *2* rebars were embedded in concrete with chlorides of 3% concentration of the cement weight for the period of 540 days. It is important that the test elements with chlorides were stored within the last year in a closed chamber over a layer of water to provide the conditions which would intensify the development of reinforcement corrosion due to high relative humidity. Fragments of the rebar in the foreground were embedded in concrete specimens. Two edge rebars *4* were protected by zinc coating, on which the products of zinc corrosion in silver colour were not particularly distinguished. Three middle black steel rebars *3* had clear signs of plentifully deposited rusty products of iron corrosion.

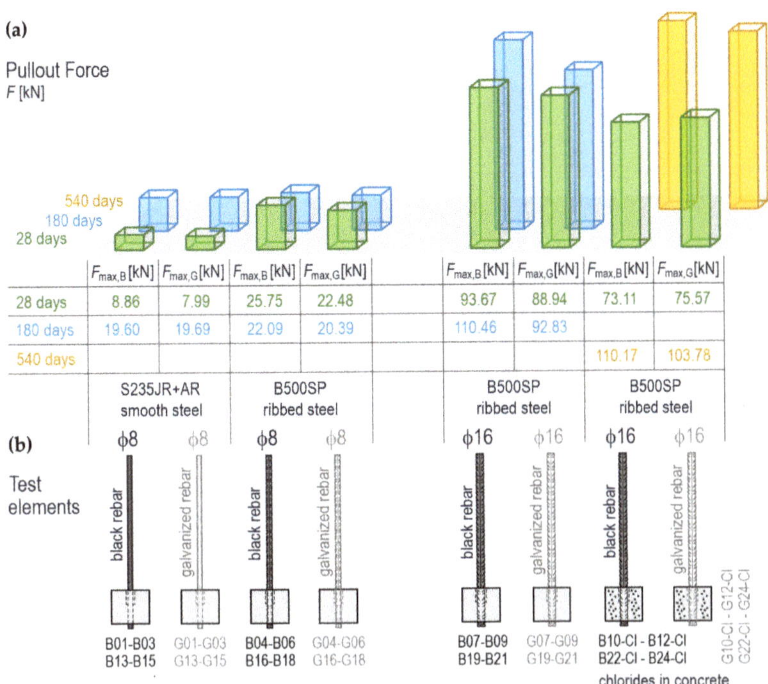

	$F_{max,B}$ [kN]	$F_{max,G}$ [kN]	$F_{max,B}$ [kN]	$F_{max,G}$ [kN]		$F_{max,B}$ [kN]	$F_{max,G}$ [kN]	$F_{max,B}$ [kN]	$F_{max,G}$ [kN]
28 days	8.86	7.99	25.75	22.48		93.67	88.94	73.11	75.57
180 days	19.60	19.69	22.09	20.39		110.46	92.83		
540 days								110.17	103.78

	S235JR+AR smooth steel		B500SP ribbed steel			B500SP ribbed steel		B500SP ribbed steel	
	ϕ8 black rebar	ϕ8 galvanized rebar	ϕ8 black rebar	ϕ8 galvanized rebar		ϕ16 black rebar	ϕ16 galvanized rebar	ϕ16 black rebar	ϕ16 galvanized rebar
	B01-B03 B13-B15	G01-G03 G13-G15	B04-B06 B16-B18	G04-G06 G16-G18		B07-B09 B19-B21	G07-G09 G19-G21	B10-CI - B12-CI B22-CI - B24-CI	G10-CI - G12-CI G22-CI - G24-CI

Figure 14. Mean values of the pull-out forces acting on galvanized and black steel (smooth ϕ8 mm and ribbed ϕ8 mm and ϕ16 mm) in the concrete specimens with and without chlorides after 28, 180, and 540 days from their preparation: (**a**) bar chart and corresponding values from the table, (**b**) the test elements; $F_{max,B}$—the pull-out force acting on black steel rebars averaged from three specimens, $F_{max,G}$—the pull-out force acting on galvanized rebars averaged from three specimens.

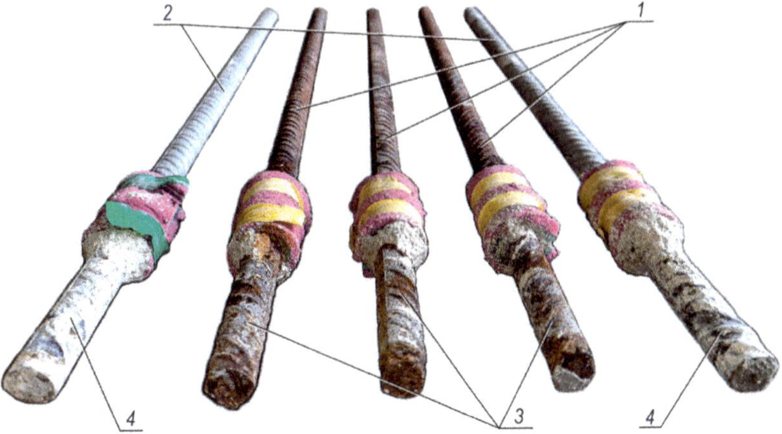

Figure 15. View of ribbed rebars ϕ16 mm made of steel grade B500SP pulled out from concrete after 540 days: *1*—black steel rebars, *2*—galvanized rebars, *3*—severely corroded fragments of black steel rebars in the area of their embedment in concrete, *4*—fragments of galvanized rebars embedded in concrete.

5. Discussion

5.1. Causes of Failure of Test Elements

Two types of damage to the test elements were observed during the strength tests. The first type consisted in pulling out the anchored rebar from the concrete specimen without breaking the cover. In that case the concrete, in which the rebar was anchored, remained whole when the ultimate resistance of the element was exceeded. All elements reinforced with rebars of 8 mm in diameter were damaged in the above way. This failure occurred for the rebars made of both steel grades S235JR+AR and B500SP. Breaking of the concrete element, in which the rebar was anchored, was the second observed type of failure. It referred to all elements reinforced with rebars $\phi16$ (B500SP). It should be emphasized that none of these elements were destroyed by the rupture of the anchored rebar.

The described types of failure of the elements are in line with the observations often described in the literature. Considering the smooth rebars, their failure was caused by pulling out the rebar from concrete, in which the rebar was anchored. The anchorage of ribbed rebars was also destroyed by pulling out the rebars from concrete providing that the reinforcement cover had adequate thickness [47]. In a situation like this, concrete was subjected to shear stress through ribs which was noticed for the rebars of 8 mm in diameter, where the cover thickness was 8.9ϕ. The cover failure occurred in the case of the rebars of 16 mm in diameter, where the cover thickness was 4.2ϕ.

To conduct a further analysis on resistance of the test elements, it is required to identify those specimens, for which the resistance failure appeared with the considerable plastic strain of the anchored reinforcement. As shown in point 2, six rebars of each diameter and steel grade were tested and the values of yielding force F_y were determined (Table A2, Appendix A). Assuming that F_y was a random variable with the standard distribution at the unknown standard deviation, then at the α significance level the F_y value was within the interval:

$$p\left(F_{y,\text{mean}} - t_\alpha \frac{s}{\sqrt{n}} < F_y < F_{y,\text{mean}} + t_\alpha \frac{s}{\sqrt{n}}\right) = 1 - \alpha \tag{1}$$

where $F_{y,\text{mean}}$—mean value of yielding force, n—number of specimens, t_α—variable with the t—student distribution for $n - 1$ degrees of freedom, s—the value estimating the standard deviation for samples according to the following relationship

$$s = \sqrt{\frac{1}{n-1} \sum_{i=1}^{n} \left(F_{y,\text{mean}} - F_{y,i}\right)^2}. \tag{2}$$

The further analyses were performed using $\alpha = 0.02$, for which $t_\alpha = 3.3649$ at the t—student distribution for $n - 1$ degrees of freedom [48]. Based on the statistical analysis of the rebars presented in Table A3 (Appendix A), the values F_y for corresponding rebars were found to be within the following ranges with the probability of $1 - \alpha = 0.98$: $\phi8$ (S235JR+AR) from 19.56 kN to 22.26 kN, $\phi8$ (B500SP) from 21.91 kN to 28.75 kN and $\phi16$ (B500SP) from 107.52 kN to 120.52 kN (Table 2).

Table 2. The results from a statistical analysis for the forces F_y yielding the test rebars.

Rebar Diameter (Steel Grade) (mm)	$F_{y,\text{mean}}$ (KN)	$F_{y,\text{mean}} - t_\alpha \frac{s}{\sqrt{n}}$ (KN)	$F_{y,\text{mean}} + t_\alpha \frac{s}{\sqrt{n}}$ (KN)	s (kN)
$\phi8$ (S235JR+AR)	20.92	19.56	22.26	0.98
$\phi8$ (B500SP)	25.33	21.91	28.75	2.49
$\phi16$ (B500SP)	114.20	107.52	120.52	4.86

In this context specifying a higher significance level, e.g., 0.05, would exclude the yielding forces that were experimentally determined, from the above ranges. Thus, a higher confidence level would lead to a contradiction. Knowing the upper and lower confidence limits for the forces F_y, the direct cause of the anchorage failure could be specified. The failure caused by plastic strain occurred when the pull-out force F reached the value F_{max} higher than the lower confidence limit for F_y, that is:

$$\frac{F_{max}}{\left(F_{y,mean} - t_\alpha \frac{s}{\sqrt{n}}\right)} > 1 \tag{3}$$

If the sign of inequality in the relationship (3) was opposite, it means the element failure could be attributed to the exceeded critical stress of bond between rebars and concrete. However, if the following relationship would be satisfied

$$\frac{F_{max}}{\left(F_{y,mean} + t_\alpha \frac{s}{\sqrt{n}}\right)} > 1 \tag{4}$$

then the failure of an element would be observed at stresses greater than the yield strength.

The values of the adequate quotients (3) and (4) are shown in Table A3, Appendix A. It follows that the relationship (4) was not satisfied for any element. Thus, the resistance of any element did not exceed the value corresponding to the force yielding the reinforcement. However, the relationship (3) was satisfied for some elements which means yielding of the reinforcement occurred. These values are indicated with the superscript letter (y) in Table 3.

Table 3. The superscript letter [y] specifying the mean pull-out forces $F_{max,B}$ (the reinforcement without coating) and $F_{max,G}$ (the galvanized reinforcement) accompanied by yielding of the rebars.

Time of Test (Day)	$F_{max,B}$ (kN)	$F_{max,G}$ (kN)	$F_{max,B}$ (kN)	$F_{max,G}$ (kN)	$F_{max,B}$ (kN)	$F_{max,G}$ (kN)	$F_{max,B}$ (kN)	$F_{max,G}$ (kN)
28	8.86	7.99	25.75 [y]	22.48 [y]	93.67	88.94	73.11	75.57
180	19.60 [y]	19.69 [y]	22.09 [y]	20.39	110.46 [y]	92.83		
540							110.17 [y]	103.78
Rebar (steel grade)	φ8 (S235JR+AR) smooth steel		φ8 (B500SP) ribbed steel		φ16 (B500SP) ribbed steel		φ16 (B500SP) ribbed steel + Cl	

5.2. Effect of Selected Parameters on Maximum Bond Stresses

Assuming that the uniform distribution of anchorage stress f_b was observed over the anchorage length l_b, the stress values could be determined from the equilibrium state of the rebar

$$f_b = \frac{F}{l_b u} \tag{5}$$

where l_b = 70 mm, u—rebar circumference. The maximum values of these stresses ($f_{b,max}$) at the resistance loss are compared in Table 4. The superscript letter (y) written for the selected values means that the failure force acting on the test element was accompanied by the strain that is associated with yielding of the rebars.

Table 4. Adhesive stresses $f_{b,max}$ of reinforcing steel to concrete at the loss of the specimen resistance.

Time of Test (Day)	$f_{b,max,B}$ (MPa)	$f_{b,max,G}$ (MPa)	$f_{b,max,B}$ (MPa)	$f_{b,max,G}$ (MPa)	$f_{b,max,B}$ (MPa)	$f_{b,max,G}$ (MPa)	$f_{b,max,B}$ (MPa)	$f_{b,max,G}$ (MPa)
28	5.04	4.54	14.64 [y]	12.78 [y]	26.62	25.82	20.78	21.48
180	11.40 [y]	19.69 [y]	12.56 [y]	11.59	31.39 [y]	26.38		
540							31.30 [y]	29.49
Rebar (steel grade)	φ8 (S235JR+AR) smooth steel		φ8 (B500SP) ribbed steel		φ16 (B500SP) ribbed steel		φ16 (B500SP) ribbed steel + Cl	

To analyse the obtained results, the bond stress $f_{b,y,mean}$ was determined in the first place which corresponded to the mean force $F_{y,mean}$ yielding the reinforcement. The values

$f_{b,y,mean}$ were: 11.89 MPa for the rebars ɸ8 S235JR+AR, 14.40 MPa for the reinforcement ɸ8, B500SP, and 32.46 MPa for the rebars ɸ8, B500SP (Table 5).

Table 5. The adhesive stress values $f_{b,y}$ corresponding to the mean force F_y yielding the reinforcement and the confidence intervals at the significance level of 0.02 corresponding to these stresses.

Rebar Diameter (Steel Grade) (mm)	$f_{b,y,mean}$ (MPa)	Minimum Value of Confidence Interval (MPa)	Maximum Value of Confidence Interval (MPa)
ɸ8 (S235JR+AR)	11.89	11.12	12.65
ɸ8 (B500SP)	14.40	12.45	16.34
ɸ16 (B500SP)	32.46	30.56	34.25

It should be pointed out that the value $f_{b,y,mean}$ for the rebar of 16 mm in diameter made of steel B500SP was more than twice the corresponding value for the same steel grade, but the diameter of 8 mm. This result was in line with the predictions because the value of adhesive stress $f_{b,y}$ corresponding to stress f_y for the reinforcement could be expressed using the rearranged expression (5):

$$f_{b,y} = \phi \frac{f_y}{4l_b}$$ (6)

Assuming that the stress $f_{b,y}$, as F_y, was the random variable, the corresponding confidence intervals at the significance level of 0.02 are additionally presented in Table 5.

The value $f_{b,y}$ was the highest possible value of stress f_b, which could be generated at the side wall of the rebar in concrete. It can be attributed to the mechanism of transmitting the force from the reinforcement to concrete. In the case of smooth rebars, where stresses f_b were caused by adhesion, the yielding force exerted on the rebar resulted in a rapid drop of its diameter, and consequently the adhesion force was eliminated due to the produced stresses perpendicular to the rebar axis. However, the force F_y initiated so significant deformation of anchored ribbed rebar that it could not be transmitted to concrete.

Table 6 contains the values expressed in percentage and determined from the equation:

$$\frac{f_{b,max}}{f_{b,y,mean}} \cdot 100\%$$ (7)

Table 6. Typical parameters determined from the graphs in Figure 12, based on the analysis of the relationship between the pull-out force F [kN] and shifts s [mm] of rebars: 0.9 F_{max} and F_{max}—90% and 100% respectively, of the maximum pull-out force, s (60 kN), s (0.9 F_{max}) and s (F_{max})—shift s of the rebar from the concrete specimen for the force of 60 kN, 90% F_{max} and F_{max}, k (0–60 kN) and k [(0.9–1.0) F_{max}]—stiffness k of the adhesive bond between the rebar and concrete for the force values within the range of 0–60 kN and 90–100% of the force F_{max}.

Specimens	0.9 F_{max} (kN)	F_{max} (kN)	s (60 kN) (mm)	s (0.9 F_{max}) (mm)	s (F_{max}) (mm)	k (0–60 kN) (kN/mm)	k (0.9–1.0) F_{max} (kN/mm)
B19	92.12	102.35	0.30	0.64	1.07	197.70	23.54
B21	107.36	119.29	0.31	1.27	1.96	193.70	17.32
G20	83.68	92.97	0.36	0.66	1.49	166.32	11.22
G21	85.43	94.92	0.36	0.72	1.37	166.34	14.57
B23-Cl	97.97	108.86	0.31	0.59	0.69	194.72	104.23
B24-Cl	100.27	111.41	0.37	0.66	0.75	164.07	130.32
G22-Cl	89.82	99.8	0.30	0.57	0.67	200.00	96.04
G24-Cl	94.17	104.63	0.37	0.49	0.56	167.07	171.85

In this table $f_{b,max}$ was considered the random variable. It means that the value determined from the Equation (7) takes 100% if f_b is within the adequate confidence level specified in Table 5. Graphical presentation of the results of calculations quotient (7) is illustrated in Figure 16. The impact of variable parameters of the test elements on the values $f_{b,max}$ was determined from the analysis of the values presented in Figure 16.

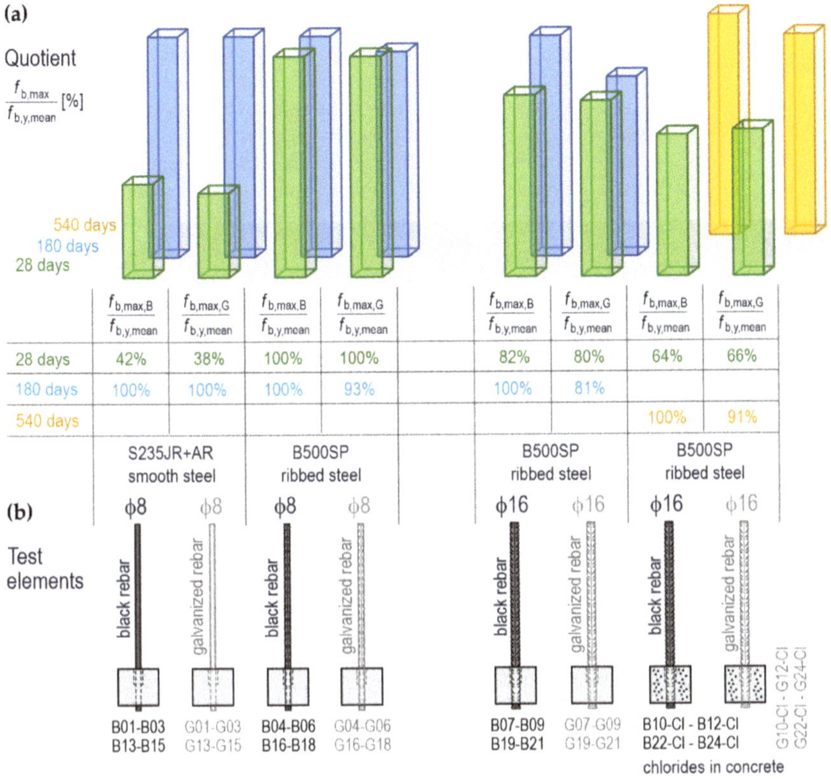

(a)

Quotient

$$\frac{f_{b,max}}{f_{b,y,mean}}\ [\%]$$

	$\dfrac{f_{b,max,B}}{f_{b,y,mean}}$	$\dfrac{f_{b,max,G}}{f_{b,y,mean}}$	$\dfrac{f_{b,max,B}}{f_{b,y,mean}}$	$\dfrac{f_{b,max,G}}{f_{b,y,mean}}$		$\dfrac{f_{b,max,B}}{f_{b,y,mean}}$	$\dfrac{f_{b,max,G}}{f_{b,y,mean}}$	$\dfrac{f_{b,max,B}}{f_{b,y,mean}}$	$\dfrac{f_{b,max,G}}{f_{b,y,mean}}$
28 days	42%	38%	100%	100%		82%	80%	64%	66%
180 days	100%	100%	100%	93%		100%	81%		
540 days								100%	91%
	S235JR+AR smooth steel		B500SP ribbed steel			B500SP ribbed steel		B500SP ribbed steel	
(b)	$\phi8$	$\phi8$	$\phi8$	$\phi8$		$\phi16$	$\phi16$	$\phi16$	$\phi16$
Test elements	black rebar	galvanized rebar	black rebar	galvanized rebar		black rebar	galvanized rebar	black rebar	galvanized rebar
	B01-B03 B13-B15	G01-G03 G13-G15	B04-B06 B16-B18	G04-G06 G16-G18		B07-B09 B19-B21	G07-G09 G19-G21	B10-CI - B12-CI B22-CI - B24-CI	G10-CI - G12-CI G22-CI - G24-CI

chlorides in concrete

Figure 16. The percentage values determined from the relationship (7) based on the assumption that the parameter f_b of adhesive stresses is the random variable: **(a)** quotient $f_{b,max}/f_{b,y,mean}$ expressed as [%], **(b)** the test elements.

Galvanization of smooth rebars at the beginning, that is, after 28 days from the day of their preparation, practically had no effect on the values $f_{b,max}$ that could be found at the rebar surface. For the galvanized smooth rebars, the values $f_{b,max}$ were lower by 4.2% compared to black steel smooth rebars. Similar observations were made for the ribbed rebars $\phi16$ mm embedded in concrete without chlorides. The values $f_{b,max}$ for galvanized rebars were lower by 2.5% than in the case of black steel rebars. Considering the rebars in concrete with chlorides, the reverse relationship was found. The values $f_{b,max}$ were greater by 2.2% after 28 days. In the case of ribbed rebars of 8 mm in diameter, the galvanization had no effect on the value $f_{b,max}$ after 28 days because the adhesive stresses in both cases reached the value $f_{b,y}$.

The values $f_{b,max}$ were increasing with the age of the test specimens which could be interpreted as a consequence of the increased strength of concrete. However, this increment was lower for the galvanized rebars with a diameter of 16 mm. The tests on the elements conducted after 180 days demonstrated that the value $f_{b,max}$ for the galvanized elements was lower by 18.7% compared to the black steel elements. A similar effect was observed for the elements with chlorides tested after 540 days. In that case, the value $f_{b,max}$ was lower by 9.1% which was not favourable to galvanized steel. A similar adverse effect of galvanization on the value f_b was found for the ribbed rebars of 8 mm in diameter that were tested after 180 days. In this case the values f_b for the galvanized rebars were lower by 9.3% compared to the black steel rebars. On the other hand, the value f_b determined for these rebars after 180 days was lower than after 28 days which was contrary to the expectations. It could indicate a serious drop in the resistance of anchorage of galvanized rebars over time. It should be mentioned that it was a single observation which could not be generalized.

The test results lead to the conclusions that chlorides in concrete had an impact on the values $f_{b,max}$. Thus, $f_{b,max}$ could be directly compared with the values determined after 28 days of concrete hardening. In the case of black steel rebars of 16 mm in diameter, chlorides reduced $f_{b,max}$ by 18%. This effect was not so significant for the galvanized rebars as the values were reduced by 13.3%. The results determined after 540 days were likely to confirm the observations that the values f_b lower by 9.1% were found in the test elements with Cl. The value $f_{b,max}$ was also affected by the process of galvanization.

5.3. Evaluation of Anchorage Stiffness of Rebars in Concrete

As described in point 4.5, the rebar shift s against the concrete specimen was recorded during the tests performed on the test elements containing the anchored rebars with a diameter of 16 mm and the measurements taken for the pull-out force F. These results are presented in the form of graphs in Figure 12. The relationships $F(s)$ showed that these graphs were linear for a gradually increasing force F within the range of 0–60 kN. However, the shape of the graph $F(s)$ was changing in the final stage of loading. For the elements without chlorides (B19, B21, G20, G21), a slope of the relationship $F(s)$ was clearly decreasing. In the case of the elements containing chlorides (B23-Cl, B24-Cl, G22-Cl, G24-Cl), a slope of the curve with a quasi-linear shape was approximately constant throughout the whole loading cycle. Two phases were identified for each path of the relationship $F(s)$ for the quantitative description of the above observations. The first phase included the loading range from 0 to 60 kN. The second phase was specified as the load contributing from 90% to 100% of the force F_{max}. Then, the shift values s were read from Table 6 for typical loading levels. The values of stiffness k of the adhesive bond between the rebar and concrete were determined for both phases. They were considered as a quotient of an increment in loading ΔF and the corresponding increase in shift Δs.

$$k = \frac{\Delta F}{\Delta s} \tag{8}$$

The calculated stiffness values are shown in Table 6 and illustrated in Figure 17. The bar chart in Figure 17a confirms the observations based on the analysis of the relationship $F(s)$.

All the elements had significant stiffness values during the first phase of loading. The tests performed after 180 days demonstrated higher stiffness k during the first loading phase for the specimens with no chlorides and with black steel elements (197.7 kN/mm and 193.7 kN/mm) compared to the stiffness of galvanized rebars (166.34 kN/mm and 193.7 kN/mm). The stiffness of the elements was significantly reduced in the final loading phase. However, higher values were still noticed for the anchorage made of black steel (23.54 kN/mm and 17.32 kN/mm) compared to the anchorage of galvanized rebars (11.22 kN/mm and 14.57 kN/mm).

The elements with chlorides that were tested after 540 days demonstrated other stiffness values. Their initial stiffness within the loading range from 0 kN to 60 kN was 164.07–200.00 kN/m. In the final loading phase that covered from 90% to 100% F_{max}, the stiffness slightly dropped to the level of 96.04–17.85 kN/m. The obtained results cannot be used to prove conclusively that galvanization of the rebars affects the stiffness of their anchorage in concrete after 540 days from the day of the preparation of these elements. However, the chlorides clearly demonstrated their impact on the stiffness in the final phase of loading. Chloride-induced corrosion of the reinforcement increased the anchorage stiffness in that phase by an order of magnitude.

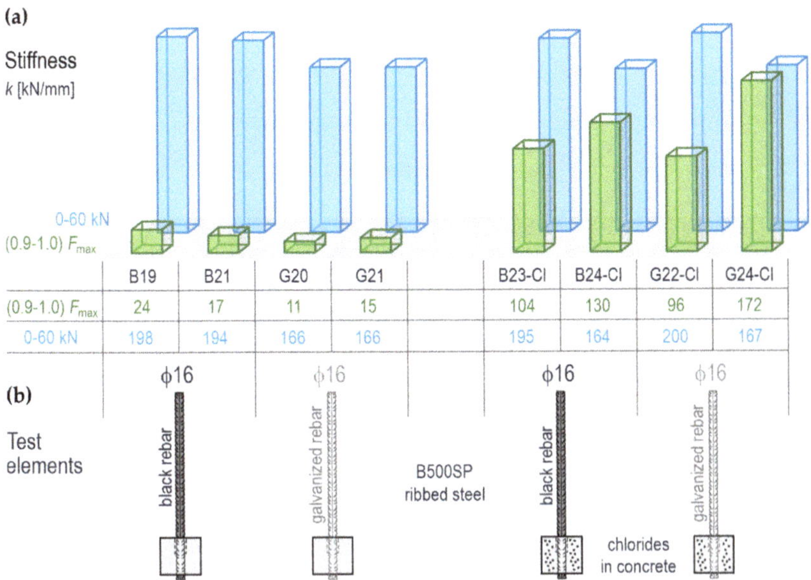

	B19	B21	G20	G21		B23-Cl	B24-Cl	G22-Cl	G24-Cl
(0.9-1.0) F_{max}	24	17	11	15		104	130	96	172
0-60 kN	198	194	166	166		195	164	200	167

Figure 17. Anchorage stiffness of ribbed (galvanized and black steel) rebars φ16 mm in concrete (with and without chlorides) determined for two loading phases: phase 1—0–60 kN, phase 2—(0.9–1.0) F_{max}: (**a**) stiffness of the test elements, (**b**) the test elements.

According to the authors, the concrete age had no significant impact on the differences determined for observed stiffness values in the final phase of loading. The mean strength of concrete f_{cm} in the test elements at the age of 180 days was 47.70 MPa, and 50.16 MPa in the case of the test elements at the age of 540 days (Table A1, Appendix A). The modulus of concrete stiffness, estimated on the basis of the standard relationship [EC2],

$$E_{cm}(t) = \left(\frac{f_{cm}(t)}{f_{cm}} \right)^{0.3} \cdot E_{cm} \tag{9}$$

increased from 35.15 GPa to 35.69 GPa (Table A1, Appendix A). This means that the products of chloride-induced corrosion, depositing in concrete pores in the immediate vicinity of the rebars, had a noticeable impact on the increased stiffness of the test bond.

6. Conclusions

The following conclusions can be drawn from testing the effect of the bond of hot-dip galvanized rebars made of steel grade B500SP and S235JR+AR to concrete and from the effects caused by chloride-induced corrosion of the reinforcement.

- The failure of all elements with ribbed rebars φ8 mm (B500SP) was accompanied by yielding of the reinforcement regardless of the age of the test elements. The presence of zinc coating on the rebars was not significant for the failure.
- Yielding of the elements reinforced with smooth rebars φ8 mm (S235JR+AR) was noticed only in the second stage of the tests, that is, after 180 days from their preparation. This observation also refers to the anchorage of galvanized and black steel rebars.
- Yielding of the reinforcement was not found in the case of failure of any test element containing galvanized ribbed rebars φ16 mm (B500SP). This fact could not be attributed to the age of the test elements or traces of products from zinc corrosion found after the pull-out of the rebars. This means that the zinc coating on the reinforcement

inhibited the full use of mechanical properties of the rebars ϕ16 mm made of steel grade B500SP.

- In the case of the elements with the anchored black steel ribbed rebars ϕ16 mm (B500SP), yielding was observed only for the elements at the age of 180 and 540 days.
- Generally, zinc coating reduced the stiffness of anchorage of ribbed rebars ϕ16 mm (B500SP) This stiffness was reduced even by 15% when the pull-out force was within the range of 0–60 kN.
- However, the chloride-induced corrosion clearly demonstrated its impact on the anchorage stiffness in the final phase of loading. For the same content of chlorides in concrete, the development of corrosion of black steel was considerably more intensive and was related to greater volume of corrosion products on the rebar surface. The products from the corrosion of iron and zinc deposited in concrete pores and filled voids, which caused an increase in the anchorage stiffness in that phase by as much as one order of magnitude.

Author Contributions: Conceptualization, M.J.; methodology, M.J.; formal analysis, M.J. and K.G.; investigation, J.K.; resources, J.K.; data curation, M.J. and J.K.; writing—original draft preparation, M.J. and K.G.; writing—review and editing, M.J. and K.G.; visualization, M.J.; supervision, M.J. All authors have read and agreed to the published version of the manuscript.

Funding: The research was financed by Silesian University of Technology (Poland) within the grant no. 03/020/RGP_19/0072 and in part within the project BK-281/RB-2/2021 (03/020/BK_21/0103).

Institutional Review Board Statement: Not applicable.

Informed Consent Statement: Not applicable.

Data Availability Statement: The data presented in this study are available on request from the corresponding author.

Conflicts of Interest: The authors declare no conflict of interest. The funders had no role in the design of the study; in the collection, analyses, or interpretation of data; in the writing of the manuscript, or in the decision to publish the results.

Appendix A

Table A1. Compressive strength of concrete in the test elements determined by testing cylindrical specimens with a diameter of 150 mm and a height of 300 mm after 28, 180, and 540 days from setting them in concrete.

Time (Day)	f_{ci} (MPa)	f_{cm} (MPa)	E_{cm} (GPa)
28	31.89		
28	43.09		
28	40.92		
28	33.95	38.22	32.9
28	33.05		
28	46.42		
180	58.49		
180	40.99		
180	41.75		
180	51.85	47.70	39.7
180	53.00		
180	40.14		
540	54.64		
540	55.03		
540	52.43		
540	41.54	50.16	42.2
540	42.24		
540	55.06		

Table A2. Strength characteristic for all types of the test rebars used in the tests, determined from the graphs of "stress σ–strain ε" illustrated in Figure 2: F_y—the force corresponding to yield strength of steel f_y, $F_{y,mean}$—the force corresponding to yield strength of steel f_y averaged from six values ε_u—steel strains at the maximum force, F_t—the force corresponding to tensile strength of steel f_t.

Rebar Diameter (Steel Grade)	Type of Steel Rebar	F_y (kN)	$F_{y,mean}$ (kN)	ε_u (%)	F_t (kN)	F_t/F_y
φ8 (S235JR+AR)	black smooth	20.02		15	24.24	1.21
φ8 (S235JR+AR)	black smooth	19.87		15	23.89	1.20
φ8 (S235JR+AR)	black smooth after pull-out	20.22		14	24.45	1.21
φ8 (S235JR+AR)	galvanized smooth	21.58	20.92	18	25.35	1.17
φ8 (S235JR+AR)	galvanized smooth	22.08		13	25.85	1.17
φ8 (S235JR+AR)	galvanized smooth after pull-out	21.73		17	26.41	1.22
φ8 (B500SP)	black ribbed	23.89		18	29.98	1.25
φ8 (B500SP)	black ribbed	23.99		19	29.78	1.24
φ8 (B500SP)	black ribbed after pull-out	25.70		8	31.29	1.22
φ8 (B500SP)	galvanized ribbed	29.02	25.33	15	32.95	1.14
φ8 (B500SP)	galvanized ribbed	22.18		12	26.21	1.18
φ8 (B500SP)	galvanized ribbed after pull-out	27.21		17	31.24	1.15
φ16 (B500SP)	black ribbed	116.78		11	139.49	1.19
φ16 (B500SP)	black ribbed	116.98		11	139.90	1.20
φ16 (B500SP)	black ribbed after pull-out	115.17		10	138.29	1.20
φ16 (B500SP)	galvanized ribbed	107.74	114.20	4	117.38	1.09
φ16 (B500SP)	galvanized ribbed	108.74		6	122.21	1.12
φ16 (B500SP)	galvanized ribbed after pull-out	119.80		11	141.10	1.18

Table A3. The values of the pull-out force F acting on galvanized and black steel rebars in concrete with and without chlorides on day 28, 180, and 540 from the day of their setting in concrete. $F_{y,mean}$—the force corresponding to yield strength of steel f_y averaged from six values. n—number of the specimens ($n = 6$), t_α—variable with the t-student distribution for $n - 1$ degrees of freedom, s—the value estimating the standard deviation for samples according to the relationship (2), F_B and F_G—the value of the pull-out force acting on galvanized and black steel rebars averaged from three measurements.

Time (Day)	Specimen No.	Rebar Diameter (Steel Grade) (mm)	F_{max} (kN)	$\dfrac{F}{(F_{y,mean}-t_\alpha \frac{s}{\sqrt{n}})}$	$\dfrac{F}{(F_{y,mean}+t_\alpha \frac{s}{\sqrt{n}})}$	$F_{max,B}$ $F_{max,G}$ (kN)
28	B01	φ8 (S235JR+AR)	8.47	0.43	0.38	
28	B02	φ8 (S235JR+AR)	6.11	0.31	0.27	8.86
28	B03	φ8 (S235JR+AR)	12.01	0.61	0.54	
28	G01	φ8 (S235JR+AR)	7.67	0.39	0.34	
28	G02	φ8 (S235JR+AR)	7.51	0.38	0.34	7.99
28	G03	φ8 (S235JR+AR)	8.79	0.45	0.39	
28	B04	φ8 (B500SP)	26.75	1.22	0.93	
28	B05	φ8 (B500SP)	25.99	1.19	0.90	25.75
28	B06	φ8 (B500SP)	24.51	1.12	0.85	
28	G04	φ8 (B500SP)	22.61	1.03	0.79	
28	G05	φ8 (B500SP)	21.56	0.98	0.75	22.48
28	G06	φ8 (B500SP)	23.27	1.06	0.81	
28	B07	φ16 (B500SP)	85.13	0.79	0.70	
28	B08	φ16 (B500SP)	98.54	0.92	0.82	93.67
28	B09	φ16 (B500SP)	97.39	0.91	0.81	
28	G07	φ16 (B500SP)	87.44	0.81	0.72	
28	G08	φ16 (B500SP)	90.88	0.85	0.75	88.94
28	G09	φ16 (B500SP)	88.49	0.82	0.73	
28	B10-Cl	φ16 (B500SP)	67.33	0.63	0.56	
28	B11-Cl	φ16 (B500SP)	73.08	0.68	0.60	73.11
28	B12-Cl	φ16 (B500SP)	78.91	0.73	0.65	
28	G10-Cl	φ16 (B500SP)	73.27	0.68	0.61	
28	G11-Cl	φ16 (B500SP)	75.26	0.70	0.62	75.57
28	G12-Cl	φ16 (B500SP)	78.18	0.73	0.65	

Table A3. *Cont.*

Time (Day)	Specimen No.	Rebar Diameter (Steel Grade) (mm)	F_{max} (kN)	$\dfrac{F}{(F_{y,mean}-t_\alpha \frac{s}{\sqrt{n}})}$	$\dfrac{F}{(F_{y,mean}+t_\alpha \frac{s}{\sqrt{n}})}$	$\dfrac{F_{max,B}}{F_{max,G}}$ (kN)
180	B13	$\phi8$ (S235JR+AR)	20.25	1.04	0.91	
180	B14	$\phi8$ (S235JR+AR)	19.03	0.97	0.85	19.60
180	B15	$\phi8$ (S235JR+AR)	19.51	1.00	0.88	
180	G13	$\phi8$ (S235JR+AR)	18.81	0.96	0.84	
180	G14	$\phi8$ (S235JR+AR)	20.11	1.03	0.90	19.69
180	G15	$\phi8$ (S235JR+AR)	20.15	1.03	0.90	
180	B16	$\phi8$ (B500SP)	21.41	0.98	0.74	
180	B17	$\phi8$ (B500SP)	22.38	1.02	0.78	22.09
180	B18	$\phi8$ (B500SP)	22.48	1.03	0.78	
180	G16	$\phi8$ (B500SP)	20.39	0.93	0.71	
180	G17	$\phi8$ (B500SP)	19.74	0.90	0.69	20.39
180	G18	$\phi8$ (B500SP)	21.04	0.96	0.73	
180	B19	$\phi16$ (B500SP)	102.35	0.95	0.85	
180	B20	$\phi16$ (B500SP)	109.74	1.02	0.91	110.46
180	B21	$\phi16$ (B500SP)	119.29	1.11	0.99	
180	G19	$\phi16$ (B500SP)	90.61	0.84	0.75	
180	G20	$\phi16$ (B500SP)	92.97	0.86	0.77	92.83
180	G21	$\phi16$ (B500SP)	94.92	0.88	0.79	
540	B22-Cl	$\phi16$ (B500SP)	109.87	1.02	0.91	
540	B23-Cl	$\phi16$ (B500SP)	108.85	1.01	0.90	110.07
540	B24-Cl	$\phi16$ (B500SP)	111.48	1.04	0.92	
540	G22-Cl	$\phi16$ (B500SP)	99.80	0.93	0.83	
540	G23-Cl	$\phi16$ (B500SP)	106.91	0.99	0.88	103.78
540	G24-Cl	$\phi16$ (B500SP)	104.63	0.97	0.87	

Appendix B

Table A4. Comparison of the results from analysing the impedance spectra of galvanized and black steel rebars ϕ 16 mm, made of steel grade B500SP, embedded in concrete with and without chlorides. The tests were conducted on days 1, 7, and 28 after setting the specimens in concrete.

Specimen No.	Δt (Day)	$E_{Ag\mid AgCl}$ (V)	R_c (Ω)	R_f (Ω)	CPE_f Y_f (Fs$^{\alpha-1}$)	α_f	R_t (Ω)	CPE_0 Y_0 (Fs$^{\alpha-1}$)	α_0	i_{corr}
B07	3	−0.097	444				2.50×10^4	1.18×10^{-2}	0.715	0.01
B08	3	−0.173	276				1.07×10^4	4.05×10^{-3}	0.587	0.02
B09	3	−0.175	290				8.68×10^4	6.54×10^{-3}	0.524	0.00
G07	3	−0.333	171				1.75×10^3	1.70×10^{-3}	0.655	0.24
G08	3	−0.292	170				2.32×10^3	1.35×10^{-3}	0.672	0.21
G09	3	−0.302	174				2.01×10^3	1.46×10^{-3}	0.664	0.23
B10-Cl	3	−0.354	26	28	9.34×10^{-3}	0.551	2.30×10^3	4.62×10^{-2}	0.625	0.06
B11-Cl	3	−0.378	24	1	7.86×10	0.029	1.13×10^3	6.86×10^{-3}	0.588	0.24
B12-Cl	3	−0.379	23	11	1.19×10^{-2}	0.526	8.88×10^3	2.32×10^{-2}	0.612	0.01
G10-Cl	3	−0.857	27	3	5.28×10^{-1}	0.099	3.41×10^1	2.23×10^{-2}	0.461	9.95
G11-Cl	3	−0.880	36	1	5.88×10^{-3}	0.671	3.44×10^1	1.43×10^{-2}	0.527	9.48
G12-Cl	3	−0.872	29	33	1.48×10^{-2}	0.619	3.28×10^1	1.78×10^{-2}	0.521	9.90
B07	10	−0.166	701				9.71×10^3	6.63×10^{-3}	0.572	0.02
B08	10	−0.238	646				6.42×10^3	3.70×10^{-3}	0.634	0.03
B09	10	−0.207	596				9.32×10^3	3.70×10^{-3}	0.598	0.02
G07	10	−0.629	538				5.06×10^3	1.69×10^{-3}	0.566	0.06
G08	10	−0.752	663				1.68×10^3	1.80×10^{-3}	0.602	0.23
G09	10	−0.713	648				4.45×10^3	1.89×10^{-3}	0.614	0.09
B10-Cl	10	−0.430	257	127	1.22×10^{-2}	0.278	4.43×10^2	1.53×10^{-1}	0.962	0.58
B11-Cl	10	−0.853	229	4949	3.04×10^{-4}	0.653	2.01×10^3	4.63×10^{-5}	0.460	0.38
B12-Cl	10	−0.427	130	151	1.01×10^{-3}	0.049	1.06×10^3	1.80×10^{-2}	0.406	0.24
G10-Cl	10	−0.686	325	89	3.71×10^{-2}	0.256	8.62×10^1	1.65×10^{-1}	0.873	2.85

Table A4. *Cont.*

Specimen No.	Δt (Day)	$E_{Ag\mid AgCl}$ (V)	R_c (Ω)	R_f (Ω)	CPE_f Y_f (Fs$^{\alpha-1}$)	α_f	R_t (Ω)	CPE_0 Y_0 (Fs$^{\alpha-1}$)	α_0	i_{corr}
G11-Cl	10	−0.778	21	2463	6.03×10^{-2}	0.328	1.10×10^2	5.43×10^{-4}	0.029	2.46
G12-Cl	10	−0.830	269	14	1.60×10^{-1}	0.744	5.91×10^1	1.45×10^{-2}	0.504	3.29
B07	25	−0.215	1066				1.40×10^5	5.29×10^{-3}	0.456	0.01
B08	25	−0.259	901				$4.79 \times 10^{+3}$	3.76×10^{-3}	0.564	0.04
B09	25	−0.238	1237				1.81×10^4	5.81×10^{-3}	0.496	0.01
G07	25	−0.684	838				3.88×10^3	1.58×10^{-3}	0.541	0.06
G08	25	−0.723	1011				3.28×10^3	1.67×10^{-3}	0.577	0.07
G09	25	−0.617	997				3.08×10^3	1.97×10^{-3}	0.507	0.08
B10-Cl	25	−0.388	400	6261	1.24×10^{-2}	0.276	9.73×10^1	7.80×10^{-5}	0.776	1.78
B11-Cl	25	−0.421	372	323	5.34×10^{-3}	0.325	9.50×10^1	4.15×10^{-5}	1.000	1.87
B12-Cl	25	−0.396	391	12	4.00×10^{-4}	0.746	3.89×10^2	1.92×10^{-2}	0.506	0.40
G10-Cl	25	−0.644	402	94	4.73×10^{-3}	0.402	1.79×10^2	3.20×10^{-3}	0.000	1.38
G11-Cl	25	−0.745	229	72750	6.30×10^{-3}	0.064	1.88×10^2	1.19×10^{-4}	0.833	1.23
G12-Cl	25	−0.429	304	1065	2.05×10^{-4}	0.397	5.13×10^2	6.29×10^{-4}	0.563	0.52

References

1. Yeih, W.; Huang, R.; Chang, J.J.; Yang, C.C. A pullout test for determining interface properties between rebar and concrete. *Adv. Cem. Based Mater.* **1997**, *5*, 57–65. [CrossRef]
2. Gao, X.; Li, N.; Ren, X. Analytic solution for the bond stress-slip relationship between rebar and concrete. *Constr. Build. Mater.* **2019**, *197*, 385–397. [CrossRef]
3. Desnerck, P.; Lees, J.M.; Morley, C.T. Bond behaviour of reinforcing bars in cracked concrete. *Constr. Build. Mater.* **2015**, *94*, 126–136. [CrossRef]
4. Dybel, P.; Kucharska, M. Development of Bond Strength of Reinforcement Steel in New Generation Concretes. *IOP Conf. Ser. Mater. Sci. Eng.* **2019**, *471*, 052058. [CrossRef]
5. Ergün, A.; Kürklü, G.; Başpinar, M.S. The effects of material properties on bond strength between reinforcing bar and concrete exposed to high temperature. *Constr. Build. Mater.* **2016**, *112*, 691–698. [CrossRef]
6. Pokorný, P.; Pernicová, R.; Tej, P.; Kolísko, J. Changes of bond strength properties of hot-dip galvanized plain bars with cement paste after 1 year of curing. *Constr. Build. Mater.* **2019**, *226*, 920–931. [CrossRef]
7. Hamad, B.S.; Jumaa, G.K. Bond strength of hot-dip galvanized hooked bars in high strength concrete structures. *Constr. Build. Mater.* **2008**, *22*, 2042–2052. [CrossRef]
8. Kayali, O.; Yeomans, S.R. Bond of ribbed galvanized reinforcing steel in concrete. *Cem. Concr. Compos.* **2000**, *22*, 459–467. [CrossRef]
9. Mianowski, K. Metoda analizy przyczepności i rys w żelbecie. *Arch. Inżynierii Lądowej* **1990**, *36*, 81–102.
10. Leibovich, O.; Yankelevsky, D.Z.; Dancygier, A.N. The effect of local rebar geometry on global strain measurements in pull out tests. *Struct. Concr.* **2020**, *21*, 413–427. [CrossRef]
11. Edwards, A.; Yannopoulos, P.T. Local bond-stress to slip relationships for hot rolled deformed bars and mild steel palin rebars. *J. ACI* **1979**, *67*, 405–420.
12. Stefanoni, M.; Angst, U.; Elsener, B. Corrosion rate of carbon steel in carbonated concrete—A critical review. *Cem. Concr. Res.* **2018**, *103*, 35–48. [CrossRef]
13. Cao, C. 3D simulation of localized steel corrosion in chloride contaminated reinforced concrete. *Constr. Build. Mater.* **2014**, *72*, 434–443. [CrossRef]
14. Stefanoni, M.; Angst, U.; Elsener, B. Local electrochemistry of reinforcement steel—Distribution of open circuit and pitting potentials on steels with different surface condition. *Corros. Sci.* **2015**, *98*, 610–618. [CrossRef]
15. Ji, Y.S.; Zhao, W.; Zhou, M.; Ma, H.R.; Zeng, P. Corrosion current distribution of macrocell and microcell of steel bar in concrete exposed to chloride environments. *Constr. Build. Mater.* **2013**, *47*, 104–110. [CrossRef]
16. Ahmad, S. Reinforcement corrosion in concrete structures, its monitoring and service life prediction—A review. *Cem. Concr. Compos.* **2003**, *25*, 459–471. [CrossRef]
17. Liu, X.; Niu, D.; Li, X.; Lv, Y.; Fu, Q. Pore solution pH for the corrosion initiation of rebars embedded in concrete under a long-term natural carbonation reaction. *Appl. Sci.* **2018**, *8*, 128. [CrossRef]
18. Jaśniok, T.; Zybura, A. The course of the polarization process during elekctrochemical tests on reinforcement corrosion. *Arch. Civ. Eng.* **2007**, *53*, 109–129.
19. Jaśniok, M.; Kołodziej, J. Testing with EIS technique to compare the effect of alkaline pH of concrete pore solution on rebars with or without zinc coating. *Ochr. Przed Korozją* **2016**, *59*, 170–174. [CrossRef]

20. Jaśniok, M.; Kołodziej, J.; Dudek, M. Assessing effects of chloride-induced corrosion of galvanized reinforcing steel in cement mortar, using impedance spectroscopy and scanning microscopy. *Ochr. przed Korozją* **2018**, *61*, 176–181. [CrossRef]

21. Dong, S.; Zhao, B.; Lin, C.; Du, R.; Hu, R.; Zhang, G.X. Corrosion behavior of epoxy/zinc duplex coated rebar embedded in concrete in ocean environment. *Constr. Build. Mater.* **2012**, *28*, 72–78. [CrossRef]

22. Kołodziej, J.; Jaśniok, M. Polarization tests concerning chloride impact on protective zinc coatings applied on reinforcing steel in curing concrete. *Ochr. Przed Korozją* **2017**, *60*, 330–334. [CrossRef]

23. Figueira, R.M.; Pereira, E.V.; Silva, C.J.R.; Salta, M.M. Corrosion protection of hot dip galvanized steel in mortar. *Port. Electrochim. Acta* **2013**, *31*, 277–287. [CrossRef]

24. Bertola, F.; Canonico, F.; Irico, S. Innovative technology constituted of sulfoaluminate cement-based concrete and non corrosive reinforcement to produce durable and sustainable reinforced concretes. In Proceedings of the 15th International Congress on the Chemistry of Cement, Prague, Czech Republic, 16–20 September 2019.

25. Dallin, G.; Gagné, M.; Goodwin, F.; Pole, S. *Continuously Galvanized Reinforcing Steel*; American Concrete Institute: Kansas City, MO, USA, 2015.

26. Sistonen, E.; Cwirzen, A.; Puttonen, J. Corrosion mechanism of hot-dip galvanised reinforcement bar in cracked concrete. *Corros. Sci.* **2008**, *50*, 3416–3428. [CrossRef]

27. Tittarelli, F.; Mobili, A.; Maria Vicerè, A.; Roventi, G.; Bellezze, T. Effect of the Type of Surface Treatment and Cement on the Chloride Induced Corrosion of Galvanized Reinforcements. *IOP Conf. Ser. Mater. Sci. Eng.* **2017**, *245*, 022088. [CrossRef]

28. *FIB Bulletin 49 Corrosion Protection of Reinforcing Steels—Technical Report 2009*; FIB: Lausanne, Switzerland, 2009.

29. Yeomans, S.R. *Galvanized Steel Reinforcement in Concrete: An Overview*; Elsevier Science: New York, NY, USA, 2004.

30. Pokorný, P.; Tej, P.; Kouřil, M. Evaluation of the impact of corrosion of hot-dip galvanized reinforcement on bond strength with concrete—A review. *Constr. Build. Mater.* **2017**, *132*, 271–298. [CrossRef]

31. Tepfers, R. Cracking of concrete cover along anchored deformed reinforcing bars. *Mag. Concr. Res.* **1979**, *31*, 3–12. [CrossRef]

32. Pereira, H.F.S.G.; Cunha, V.M.C.F.; Sena-Cruz, J. Numerical simulation of galvanized rebars pullout. *Frat. Integrita Strutt.* **2015**, *31*, 54–66. [CrossRef]

33. Goto, Y. Cracks formed in concrete around deformed tension bars. *ACI J.* **1971**, *68*, 244–251.

34. Lin, H.; Zhao, Y.; Feng, P.; Ye, H.; Ozbolt, J.; Jiang, C.; Yang, J.Q. State-of-the-art review on the bond properties of corroded reinforcing steel bar. *Constr. Build. Mater.* **2019**, *213*, 216–233. [CrossRef]

35. Blomfors, M.; Zandi, K.; Lundgren, K.; Coronelli, D. Engineering bond model for corroded reinforcement. *Eng. Struct.* **2018**, *156*, 394–410. [CrossRef]

36. Chinn, J.; Ferguson, P.M.; Thompson, J.N. Lapped splices in reinforced concrete beams. *J. Am. Concr. Inst.* **1955**, *27*, 201–214.

37. Manfredi, G.; Pecce, M. Behaviour of bond between concrete and steel in a large post-yelding field. *Mater. Struct. Matériaux Constr.* **1996**, *29*, 506–513. [CrossRef]

38. Jaśniok, M.; Sozańska, M.; Kołodziej, J.; Chmiela, B. A Two-Year Evaluation of Corrosion-Induced Damage to Hot Galvanized Reinforcing Steel B500SP in Chloride Contaminated Concrete. *Materials* **2020**, *13*, 3315. [CrossRef]

39. Tepfers, R. Lapped tensile reinforcement splices. *J. Struct. Div. Proc. Am. Soc. Civ. Eng.* **1982**, *108*, 283–301.

40. Law, D.W.; Molyneaux, T.C.K. Impact of corrosion on bond in uncracked concrete with confined and unconfined rebar. *Constr. Build. Mater.* **2017**, *155*, 550–559. [CrossRef]

41. Li, F.; Yuan, Y. Effects of corrosion on bond behavior between steel strand and concrete. *Constr. Build. Mater.* **2013**, *38*, 413–422. [CrossRef]

42. Coccia, S.; Imperatore, S.; Rinaldi, Z. Influence of corrosion on the bond strength of steel rebars in concrete. *Mater. Struct. Constr.* **2016**, *49*, 537–551. [CrossRef]

43. Lin, H.; Zhao, Y.; Yang, J.Q.; Feng, P.; Ozbolt, J.; Ye, H. Effects of the corrosion of main bar and stirrups on the bond behavior of reinforcing steel bar. *Constr. Build. Mater.* **2019**, *225*, 13–28. [CrossRef]

44. Bilcik, J.; Holly, I. Effect of reinforcement corrosion on bond behaviour. *Procedia Eng.* **2013**, *65*, 248–253. [CrossRef]

45. *EN 1766 Eurocode: Products and Systems for the Protection and Repair of Concrete Structures—Test Methods—Reference Concretes for Testing*; CEN-CENELEC: Brussels, Belgium, 2017.

46. *EN 1992-1-1. Eurocode 2: Design of Concrete Structures-Part 1-1: General Rules and Rules for Buildings*; CEN-CENELEC: Brussels, Belgium, 2008.

47. Ferguson, P.; Thompson, J. Development length of high strength reinforcing bars in bond. *J. Am. Concr. Inst.* **1962**, *59*, 887–992.

48. Volk, W. *Applied Statistics for Engineers*; McGraw-Hill, Inc.: New York, NY, USA, 1969.

Article

Effect of Cathodic Protection on Reinforced Concrete with Fly Ash Using Electrochemical Noise

Jorge García-Contreras [1], Citlalli Gaona-Tiburcio [2,*], Irene López-Cazares [3], Guillermo Sanchéz-Díaz [1], Juan Carlos Ibarra Castillo [1], Jesús Jáquez-Muñoz [2], Demetrio Nieves-Mendoza [4], Erick Maldonado-Bandala [4], Javier Olguín-Coca [5], Luis Daimir López-León [5] and Facundo Almeraya-Calderón [2,*]

[1] Universidad Autonoma de San Luis de Potosi, Facultad de Ingeniería, Av. Dr. Manuel Nava #8 Zona Universitaria, 78290 San Luis Potosí, Mexico; jorge.garcia@uaslp.mx (J.G.-C.); guillermo.sanchez@uaslp.mx (G.S.-D.); carlos.castillo@uaslp.mx (J.C.I.C.)

[2] Universidad Autonoma de Nuevo Leon, FIME-Centro de Investigación e Innovación en Ingeniería Aeronáutica (CIIIA), Av., Universidad s/n, Ciudad Universitaria, 66455 San Nicolás de los Garza, Mexico; Jesus.jaquezmn@uanl.edu.mx

[3] Instituto Potosino de Investigación Científica y Tecnológica, IPICYT, División de Ciencias Ambientales, Camino a la Presa de San José 2055, Lomas 4 Sección, C.P., 78216 San Luis Potosí, Mexico; irene.lopez@ipicyt.edu.mx

[4] Universidad Veracruzana, Facultad de Ingeniería Civil, 91000 Xalapa, Mexico; dnieves@uv.mx (D.N.-M.); erimaldonado@uv.mx (E.M.-B.)

[5] Universidad Autónoma del Estado de Hidalgo, Área Académica de Ingeniería y Arquitectura, Carretera Pachuca-Tulancingo Km. 4.5., 42082 Hidalgo, Mexico; olguinc@uaeh.edu.mx (J.O.-C.); luis_lopez@uaeh.edu.mx (L.D.L.-L.)

* Correspondence: citlalli.gaonatbr@uanl.edu.mx (C.G.-T.); facundo.almerayacld@uanl.edu.mx (F.A.-C.)

Citation: García-Contreras, J.; Gaona-Tiburcio, C.; López-Cazares, I.; Sanchéz-Díaz, G.; Ibarra Castillo, J.C.; Jáquez-Muñoz, J.; Nieves-Mendoza, D.; Maldonado-Bandala, E.; Olguín-Coca, J.; López-León, L.D.; et al. Effect of Cathodic Protection on Reinforced Concrete with Fly Ash Using Electrochemical Noise. *Materials* **2021**, *14*, 2438. https://doi.org/10.3390/ma14092438

Academic Editor: Luigi Coppola

Received: 10 April 2021
Accepted: 28 April 2021
Published: 7 May 2021

Publisher's Note: MDPI stays neutral with regard to jurisdictional claims in published maps and institutional affiliations.

Abstract: Corrosion of steel reinforcement is the major factor that limits the durability and serviceability performance of reinforced concrete structures. Impressed current cathodic protection (ICCP) is a widely used method to protect steel reinforcements against corrosion. This research aimed to study the effect of cathodic protection on reinforced concrete with fly ash using electrochemical noise (EN). Two types of reinforced concrete mixtures were manufactured; 100% Ordinary Portland Cement (OCP) and replacing 15% of cement using fly ash (OCPFA). The specimens were under-designed protected conditions ($-1000 \leq E \leq -850$ mV vs. Ag/AgCl) and cathodic overprotection ($E < -1000$ mV vs. Ag/AgCl) by impressed current, and specimens concrete were immersed in a 3.5 wt.% sodium chloride (NaCl) Solution. The analysis of electrochemical noise-time series showed that the mixtures microstructure influenced the corrosion process. Transients of uniform corrosion were observed in the specimens elaborated with (OPC), unlike those elaborated with (OPCFA). This phenomenon marked the difference in the concrete matrix's hydration products, preventing Cl^- ions flow and showing passive current and potential transients in most specimens.

Keywords: electrochemical noise; fly ash; concrete; cathodic protection; microstructure

1. Introduction

In the construction industry, corrosion of steel reinforcement is one of the main issues. It is considered the leading cause of the premature deterioration of reinforced concrete structures [1–3]. The most predominant factors affecting the steel bars can be physical, mechanical, biological, or chemical [4,5]. Corrosion is the most frequent and relevant deterioration suffered by reinforced concrete structures, particularly in structures located in an aggressive marine environment [6,7]. Corrosion of reinforcement in the marine environment usually occurs because aggressive agents such as chloride ions are an electrochemical process [8,9].

The cement industry is one of the two largest producers of carbon dioxide (CO_2), generating between 5–8% CO_2 total emissions to the environment. In the future, the

109

CO_2 emissions are predicted around 10–15% [2,10]. The supplementary cementitious materials (SCM) are employed to reduce CO_2 emissions and replace, in a portion, Portland cement [2,11–15].

New alkali-activated materials have been employed in place of OPC with a different type of ashes. The employ of different ashes showed good results increasing the durability of concrete. OPC has been replaced by ashes as slags furnace ash (SFA), metakaolin (MK), and fly ash (FA), among others [16–18]. SCM as sugar cane bagasse ash (SCBA) and rice husks ash (RHA) has been studied in the last 20 years to develop a better solution for reinforced structure [19–22]. Diverse research has been reported on the corrosion behavior of SCM and the structural, mechanical, chemical and structural properties [23–25]. The employ of SCM is an eco-friendly and cost-effective solution due to the natural properties of these materials [8,26]. An essential factor of SCM materials is their faster cure time in comparison with OPC.

Another critical factor of these SCM is that they cure faster than OPC, making them even more suitable for precast components.

Corrosion of reinforcing steel is the leading cause of deterioration in concrete. When steel corrodes, the resulting rust occupies a greater volume than the steel. The action of Cl^- ions are considered the most critical factor influencing the corrosion intensity of reinforced concrete [27–29]. The Cl^- ion causes the breakdown of the steel's normal passive condition in concrete and corrosion development. An efficient method to control steel reinforcement corrosion in concrete structures is the impressed current cathodic protection (ICCP). ICCP is an efficient method to stop or control the steel reinforcement's corrosion in concrete structures. ICCP is known for active or passive steel conditions; it is necessary to have specific conditions for both pH and potential in the electrochemical system [27–33]. On the other hand, studies reported that the fly ash particles react with the calcium hydroxide, generating hydration products that strongly affect decreasing the concretes porosity. Also, mechanical properties are increased when using these ashes and blast furnace slag, given the increasing volume of C-S-H gel [32].

Electrochemical noise (EN) technique for corrosion applications has allowed many advances in recent years interesting for corrosion science. A particular advantage of EN measurements is detecting and analyzing the early stages of localized corrosion.

Electrochemical noise describes the spontaneous low-level potential and current fluctuations occurring during a corrosion process. During the corrosion process, predominantly electrochemical cathodic and anodic reactions can cause small transients in electrical charges on the electrode. These transients manifest in potential and current noise exploited in a corrosion map [34,35]. The rupture and re-passivation of passive films are associated with the stochastic process (anodic and cathodic transients), as well as the formation and pitting propagation are related to the deterministic process [36,37].

Transients are linked to anodic and cathodic reactions because of stochastic processes (rupture and re-passivation of the passive film) and deterministic processes (formation and propagation of pitting) [36,37]. The time-series of potential or current provides helpful information about the corrosion process.

Ma, et al. [38,39] indicate that EN data is influenced by the measurement mode, the surface area of the working electrodes, the electrolytic resistance and the symmetry of the electrode system. Xia, et al. [40] mention several mathematical methods and parameters that can analyze EN data to identify corrosion form and corrosion rates. They are classified into three groups: the time domain, the frequency domain, and the time-frequency domain. The statistical analysis it includes parameters as noise resistance (Rn), Skewness, Kurtosis, localization index (*LI*) Chaos analysis, Recurrence quantification analysis and Fractal analysis. The Fast Fourier Transform includes power spectral density, noise impedance, etc; and time-frequency domains method; have the analysis Hilbert-Huang transform, Discrete Wavelet transform, Stockwell transform and others) [41–47]. *LI*, skewness, and kurtosis values have been related to different corrosion types and the asymmetry of the distribution of EN data [47–52].

This research aimed to study the Effect of cathodic protection on reinforced concrete with fly ash using electrochemical noise. Two types of reinforced concrete mixtures were elaborated: 100% Ordinary Portland Cement (OCP) and replacing 15% of cement using fly ash (OCPFA). The specimens were under-designed protected conditions ($-1000 \leq E \leq -850$ mV vs. Ag/AgCl) and cathodic overprotection ($E < -1000$ mV vs. Ag/AgCl) by impressed current, and specimens concrete were immersed in a 3.5 wt.% sodium chloride (NaCl) Solution. This concentration was used to simulate an aggressive environment that contains chlorides where civil works can be built based on reinforced concrete.

2. Materials and Methods

2.1. Reinforced Concrete Specimens

This research were designed for eigth specimens: OCP and FA concretes mixtures according to NMX-C-414-ONNCCE [53] of 30 MPa at 28 days. The Fly Ash (FA)—Class F type addition was according to ASTM C618-94a [54], obtained from a coal-fired power plant in Coahuila, Mexico. The Research group has tested supplementary materials as a partial substitute cement and has found good results with fly ash [2,12,19,22,23,27]. Therefore, in this study Fly Ash was used as a partial substitute for OPC with a replacement percentage of 15%. Table 1 shows the chemical composition of both materials used obtained by X-ray fluorescence (XRF) analysis provided by the supplier.

Table 1. Chemical composition of OPC and FA obtained by XRF.

Material	Compounds (wt.%)									
	CaO	SiO$_2$	Al$_2$O$_3$	Fe$_2$O$_3$	MgO	SO$_4^{2-}$	Na$_2$O	K$_2$O	MnO	TiO$_2$
Ordinary Portland Cement (CPC)	65.31	18.47	4.13	3.80	1.42	4.64	0.46	1.13	0.19	0.29
Fly Ash (FA)	3.26	57.3	28.14	5.21	0.56	0.32	0.51	1.52	-	1.21

The dosage of concrete mixtures was carried out according to the method of ACI 211.1 [55]. This method is based on the physical properties of coarse and fine aggregates, see Table 2.

Table 2. Dosage of conventional concrete and sustainable concrete, (kg/m^3 of concrete, ratio w/c = 0.66).

Mixture	Fly Ash kg/m^3	Aggregate Coarse kg/m^3	Aggregate Fine kg/m^3	Cement kg/m^3	Water kg/m^3
OPC	-	1049	781	310	205
OPCFA	46.5	1049	781	263.5	205

The OPC and FA mixtures were made with a ratio of water/cement of 0.66. The specimens were cylindrical with dimensions of 15 high by 7.5 cm in diameter [53]. In all the specimens, AISI 1018 Carbon Steel bars were embedded. The steel bars dimensions were 9.5 mm in diameter and 10.5 cm in length. The curing of all specimens was carried out by immersion in water for 28 days, according to NMX-C-159 standard [56].

2.2. System of Protection and Cathodic Overprotection

After the curing step, the specimens of both mixtures were immersed in a 3.5 wt.% NaCl solution and an impressed current cathodic protection system were induced. Anode, $\frac{1}{4}$ in. graphite bars with 0.635 cm. in diameter and a length of 15 cm. were used. The capacity of the employed rectifier was 20 V/10 A. Four of the specimens were treated to a potential protection criterion (-1000 mV $\leq E \leq -850$ mV vs. Ag/AgCl), and the rest of them with overprotection ($E < -1000$ mV vs. Ag/AgCl) [7]. The protection and overprotection levels were consistently maintained by an experimental system, as shown in Figure 1. The nomenclature used considers the following: 2–3 = months of exposure, FA = Fly ash, P = Protected, OP = overprotected.

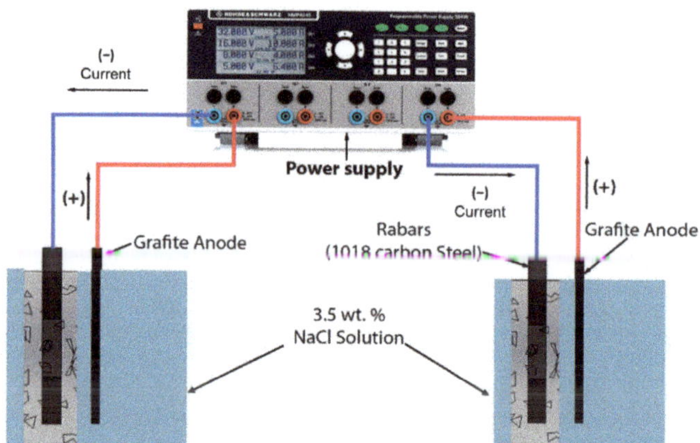

Figure 1. Experimental system to maintain the specimens under protection and cathodic overprotection.

2.3. Electrochemical Noise Test

The electrochemical noise (EN) technique was used to evaluate 100% Ordinary Portland Cement (OCP) and the other with replacing 15% of cement using fly ash (OCPFA), immersed in 3.5 wt.% in NaCl solutions. The EN measurements were carried out using ACM Instruments (Manchester, UK)—Gill-AC potentiostat/galvanostat/ZRA (Zero Resistance Ammeter).

The EN experiments were carried out in based to ASTM G199-09 standard to determine the noise resistance (R_n) and corrosion rate. To each experiment was employed two nominally identical specimens as the working electrodes (WE1 and WE2) and a saturated calomel electrode as the reference electrode (RE) [51,57]. The electrochemical current noise (ECN) was measured with a galvanic coupling current between two identical working electrodes. The electrochemical potential noise (EPN) was measured linking one of the working electrodes and the reference electrode. The current and potential electrochemical noise was monitored concerning each electrode's electrolyte combination under open-circuit conditions. 1024 data points were obtained with a scanning rate of 1 data per second. The current and potential electrochemical noise was monitored concerning each electrode's electrolyte combination under open-circuit conditions. 1024 data points were obtained with a scanning rate of 1 data per second [51,52].

DC trend signal was removed from the original EN signal by the polynomial method, from signal without DC statistical data (R_n, Kurtosis, and skewness) was obtained from the signal without DC. For PSD (power spectral density) analysis, a Hann window was applied before being transformed to the FFT frequency domain (fast Fourier transform). Data analysis was processed with a program made in MATLAB 2018a software (Math Works, Natick, MA, USA).

2.4. Scanning Electron Microscopy (SEM)

Scanning electron microscopy (SEM, Jeol JSM 5610LV, Tokyo, Japan) was as well utilized for concrete surface morphology analyzed to different magnifications, operating at 20 kV, WD = 10 mm. The chemical composition of these concrete specimens was obtained by energy dispersive X-ray spectroscopy (EDS, EDAX, Tokyo, Japan).

3. Results

3.1. Microstructure of OPC and OPCFA by Scanning Electron Microscopy.

Figure 2 shows the microstructure of specimens made with the mixture of OPC. In the 2OPC specimen (Figure 2a), it is possible to visualize the hydration products in the form of cotton. However, the elemental analysis by X-ray dispersive energy spectroscopy (EDS) shows that the Ca/Si ratio is 2.24, which is within the range of a concrete porous [58,59].

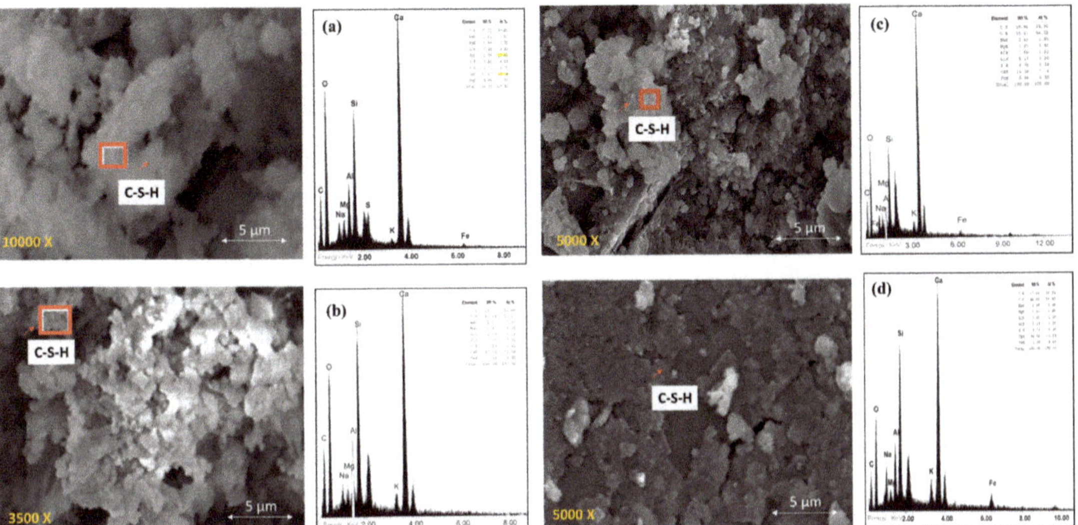

Figure 2. SEM surface morphology micrographs and elemental analysis by X-ray dispersive energy spectroscopy (EDS) of specimens made with the OPC mixture: (**a**) 2OPC-P Ca/Si = 2.24, (**b**) 2OPC-OP Ca/Si = 1.86, (**c**) 3OPC-P Ca/Si = 2.23, (**d**) 3OPC-OP Ca/Si = 3.25.

The specimen 2OPC-OP (Figure 2b) shows that the concrete obtained higher hydration than protected. The hydration occurs because the Ca/Si ratio is 1.86, which is in the range of well-hydrated concrete. According to the EDS analysis, a higher amount of hydrated calcium silicates (C-S-H gel) [27,32,60].

For specimens with three months of exposure (Figure 2c,d), it is observed that the microstructure of the concrete is more porous than those exposed to two months. Porous could be attributed to Ca/Si ratios in intervals of 2.23 and 3.25, which is typical of porous concrete. With very few hydration products (C-S-H gel) shown by the images and reinforced by the EDS analysis.

Figure 3 shows the microstructure of the specimens made with the mixture with OPCFA. In Figure 3a,b, the microstructure of the specimen was exposed to two months under protection and overprotection conditions. The 2OPCFA-P specimen shows an appreciable amount of hydration products (C-S-H gel). According to the EDS analysis, the Ca/Si ratio is within a hydrated concrete range with a value of 1.78. As for specimen 2OPCFA-OP, it is observed that the Ca/Si ratio increased to a value of 2.45, indicating greater porosity but with fewer hydration products in this specimen. In the specimens at three months of exposure with both protective conditions, they show the effects of fly ash particles, generating more hydration products. In Figure 3c,d reinforced by the EDS analysis with the Ca/Si ratios results, which tends to decrease with time in this type of mixtures, depending on the hydration time and the amount of fly ash in the mixture.

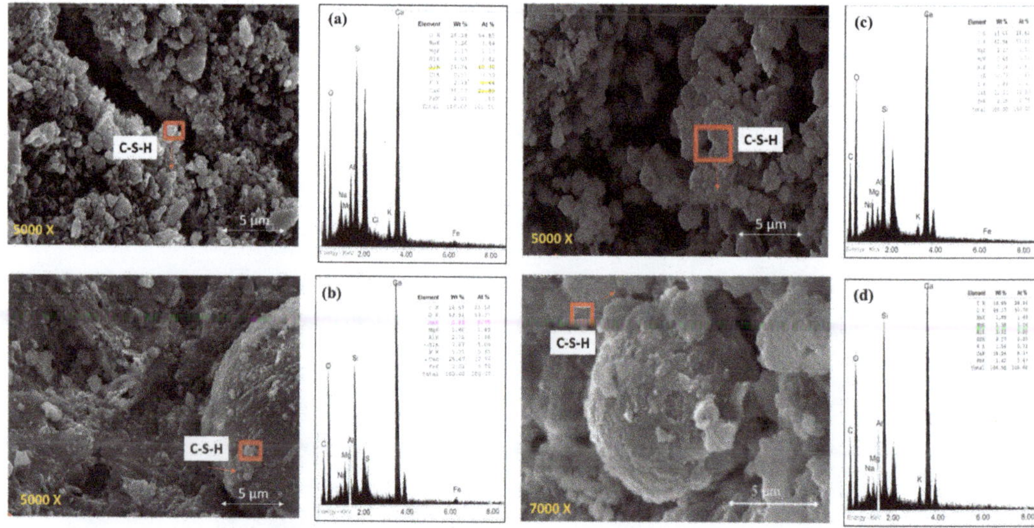

Figure 3. SEM surface morphology micrographs and elemental analysis by X-ray dispersive energy spectroscopy (EDS) of specimens made with the OPCFA mixture: (**a**) 2OPCFA-P Ca/Si = 1.78, (**b**) 2OPCFA-OP Ca/Si = 2.45, (**c**) 3OPCFA-P Ca/Si = 1.44, (**d**) 3OPCFA-OP Ca/Si = 1.40.

3.2. Electrochemical Noise

EN signal is composed of random, stationary, and DC variables. It is necessary to separate DC from random and stationary components to analyze EN data because DC creates false frequencies and interferes in visual, statistical, and PSD analysis. In this way, when DC is removed, corrosion data presented at low frequencies are conserved [51]. EN can be represented by Equation (1) [43–45]:

$$x(t) = m_t + s_t + Y_t \tag{1}$$

where $x(t)$ is the EN time series, m_t, is the DC component, s_t is the random component, and Y_t is a stationary component. The latter two are the functions that define the corrosion system [36]. The polynomial method, as defined in Equation (2), defines noise signal (x_n) and polynomial of "n" grade (p_o) at n-th term (a_i) in "n" time to obtain a signal without trend (y_n) [33,42–44]:

$$y_n = x_n - \sum_{i=0}^{p_o} a_i n^i \tag{2}$$

Figure 4 shows the EN signal for specimens subjected to cathodic protection current. The EPN signal of OPC-P samples exposure two months (Figure 4a,b) presents potential transients of 3×10^{-4} V maximum. Low potentials are associated with high stability produced by oxidation-reduction events. The ECN signal corroborated it with low amplitude transients and fluctuations of $\times 10^{-7}$ (A/cm^2) order. The behavior of EPN and ECN signals is related to the uniform corrosion process. In the case of FA samples, the potential transients show individual events of low frequency. On the other hand, OPC-P samples presented a higher instability associated with the passive layer's break. When concrete contains FA and the passive layer is broken, it will be regenerated easier than samples with only OPC.

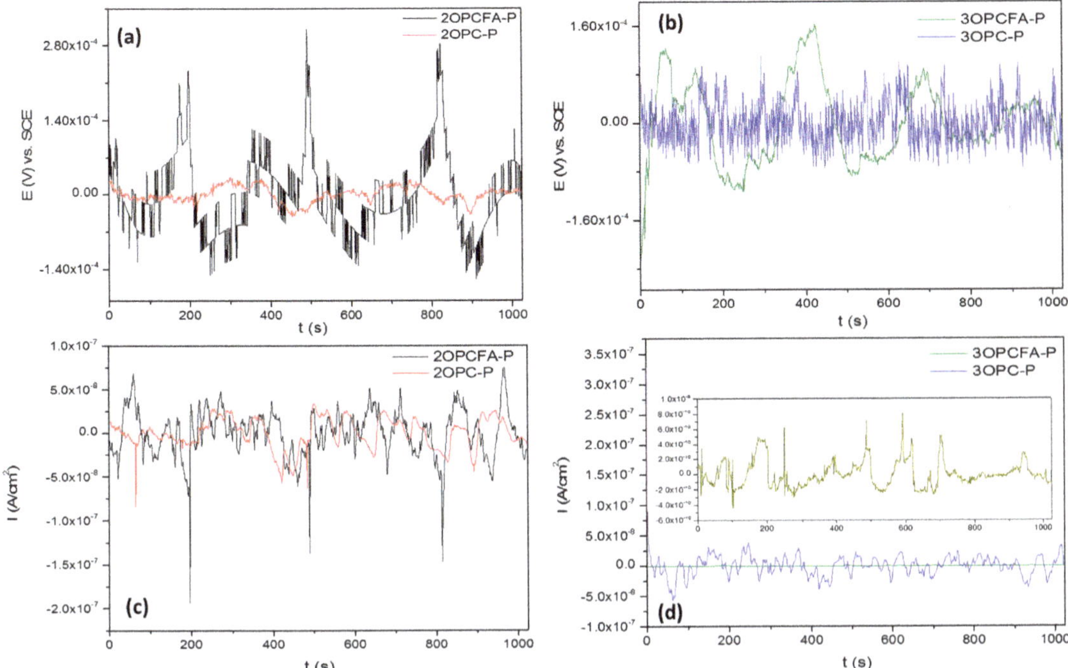

Figure 4. Electrochemical noise-time series for specimens subjected to Impressed current cathodic protection of reinforced concrete mixtures (OPC-P and OPCFA-P). (**a**,**b**) ENP, 2 and 3 months, (**c**,**d**) ENC, 2 and 3 months.

The sample of OPC-P exposed three months (Figure 4c,d) presented similar behaviors in EPN and ECN signals, associated with uniform corrosion. However, the sample with FA presented current values of $\times 10^{-10}$ (A/cm^2) amplitude. The low current value indicates passivation in the sample after three months of exposure [13]. The decrease of current demand suggests that the passive layers become more stable with the exposure. Also, at three months, EPN and ECN signals did not present the anodic transients that samples with two months of exposure presented.

Figure 5 shows the EN signal of samples subjected to cathodic overprotection current. For overprotected samples, the current and potential values are low frequencies and amplitudes. However, the variation of fluctuations is high, associated with an unstable system. The FA samples showed the same behavior, with low amplitude and signals but with high variations; this could provoke the system's instability. Also, samples with the only OPC presented higher current demand indicating a higher corrosion kinetic. The sample 2OPCFA-OP presented potential and current transients of very short frequencies and very low amplitudes. This behavior was presented due to stable passivation and ionic diffusion through this passive film.

The three months samples with OPC presented potential and current transients characteristic of unstable uniform corrosion, with low frequency and low amplitude behaviors. However, the decrease of frequency presented at two months of exposure is related to a higher harmful activity. The samples with FA presented lower current demand related to lower corrosion kinetic and suggest a passivity phenomenon.

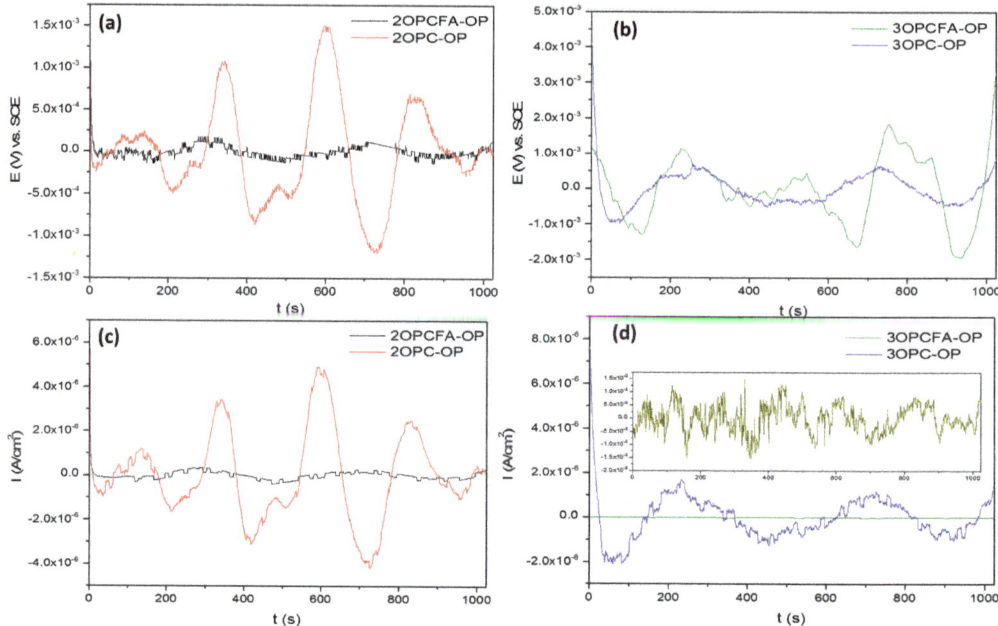

Figure 5. Electrochemical noise-time series for specimens subjected to Impressed current cathodic protection of reinforced concrete mixtures (OPC-OP and OPCFA-OP). (**a,b**) ENP, 2 and 3 months, (**c,d**) ENC, 2 and 3 months.

3.2.1. Statistical Analysis

Electrochemical noise resistance (R_n) is determined by standard deviation from time series values. These statistical values give corrosion kinetics information. Cottis and Turgoose [50] found a relationship between the increase of variance and standard deviation with an increase in corrosion rate. The standard deviation (S) of the current or potential is calculated using the relationship.

$$\sigma_x = \sqrt{\overline{x^2}} = \sqrt{\frac{1}{N}\sum_{i=1}^{N}(x_1 - \overline{x})^2} \tag{3}$$

where x_1 is the transient ECN or transient EPN, $\overline{x}$ the average ECN or average EPN, and n is the number of pints in the recording.

Noise resistance (R_n) is defined as the ratio of the standard deviation of potential to the standard deviation of current: (Equation (4)).

$$R_n = \frac{\sigma_v}{\sigma_I} * A \tag{4}$$

R_n and R_p are equivalent to the Stern–Geary equation (Equation (5) can be applied as an analog relation to determining corrosion. R_p is expressed in $\Omega \cdot cm^2$, i_{corr} is the corrosion current density, and B in V. is a constant resulting from a combination of the anodic and cathodic Tafel slopes. Recommended values of B are 0.026 and 0.052 V for active and passive systems, respectively [61].

$$R_n = \frac{B}{i_{corr}} = \frac{b_a b_c}{2.303\,(b_a + b_c)\cdot i_{corr}} \tag{5}$$

Some authors related the EN signal to statistical analysis with the metal surface to the corrosion process [51]. The $I_{r.m.s}$ is obtained by Equation (6), where X is the average of EN data, n the data number, and σ the standard deviation:

$$r.m.s = \sqrt{Xn^2 + \sigma^2} \tag{6}$$

The localization index (LI) is defined as the ratio of the standard deviation of ECN, σ_i, and $I_{r.m.s}$ (root mean square of the noise current) (Equation (7)).

$$LI = \frac{\sigma_i}{I_{r.m.s}} \tag{7}$$

Values obtained can be associated with the system is corrosion type [51,52,62]. This research will take more parameters to determine the corrosion type (Table 3).

Table 3. Correlation between LI and type of corrosion expected [60].

Corrosion Type	LI (Value Range)
Localized	1.0–0.1
Mixed	0.1–0.01
Uniform	0.01–0.001

This research employed Kurtosis and skewness to try to define the corrosion type. Localization index (LI) was not considered because Mansfeld and Sun [63] in 1995 concluded that LI can present limitations and should be used with discretion. In 2001, Reid and Eden [64] developed a patent where they identified corrosion type based on statistical moments with skewness and Kurtosis (Equations (8) and (9)), which are the 3rd and 4th statistical moments [51,63,64]:

$$skewness = \frac{1}{N} \sum_{i=1}^{N} \frac{(x_i - \bar{x})^3}{\sigma^3} \tag{8}$$

$$kurtosis = \frac{1}{N} \sum_{i=1}^{N} \frac{(x_i - \bar{x})^4}{\sigma^4} \tag{9}$$

Statistical calculations have a standard error (SE) that generates uncertainty in the results. The following Equation can be provided, where N is the number of data studied [65]. Hence, when the data number is significant, the standard error will be lower than when the data number is high.

$$SE = \sqrt{\frac{24}{N}} \tag{10}$$

SE is 0.153; values obtained will take SE as a parameter of uncertainty. Corrosion type determined by Kurtosis and skewness is shown in Table 4:

Table 4. Corrosion types evaluated by Kurtosis and skewness [65].

Corrosion Type	Potential		Current	
	Skewness	Kurtosis	Skewness	Kurtosis
Uniform	<±1	<3	<±1	<3
Pitting	<−2	>>3	>±2	>>3
Transgranular (SCC)	4	20	−4	20
Intergranular (SCC #1)	−6.6	18 to 114	1.5 to 3.2	6.4 to 15.6
Intergranular (SCC #2)	−2 to −6	5 to 45	3 to 6	10 to 60

Kurtosis and skewness were applied to the ECN signal to determine the corrosion mechanism based on corrosion kinetic. Table 5 shows R_n, i_{corr}, skewness, and Kurtosis from EN signal filtered with a 9th-grade polynomial to remove DC signal. For protected samples, *LI* indicates values near to 1; this value is related to a disordered system and high variation of standard deviation. For that reason, it is essential to analyze with more precise methods as Skewness and Kurtosis in potential. Employing skewness and analyze the protected samples presented uniform corrosion.

Table 5. EN statistical parameters of specimens subjected to Impressed current cathodic protection of reinforced concrete mixtures (OPC-P/OP and OPCFA-P/OP).

Sample	R_n ($\Omega \cdot cm^2$)	i_{corr} (mA/cm^2)	*LI*	Corrosion Type	Skewness (Pot)	Corrosion Type	Kurtosis (Pot)	Corrosion Type
				Protected				
2OPCFA-P	2779.05	0.0095557	1	Localized	−0.82	Uniform	6.7	Pitting
2OPC-P	826.78	0.31447	1	Localized	−0.65	Uniform	2.96	Pitting
3OPCFA-P	383777.68	0.0000677	1	Localized	0.19	Uniform	2.89	Uniform
3OPC-P	1830.34	0.0142	1	Localized	0.38	Uniform	2.81	Uniform
				Overprotected				
2OPCFA-OP	404.08	0.0643422	0.49	Localized	−1.77	Pitting	14.24	Pitting
2OPC-OP	296.74	0.876967	0.93	Localized	0.46	Uniform	3.11	Pitting
3OPCFA-OP	177951.805	0.0001461	1	Localized	0.09	Uniform	2.93	Uniform
3OPC-OP	493.7	0.0526635	1	Localized	1.9	Pitting	14	Pitting

The overprotected samples presented high values of *LI*, skewness, and Kurtosis. High *LI*, skewness, and Kurtosis values are associated with the high current variation with low current and potential values; any variation can change the results. The samples with FA presented minor variation; this could be attributed to a passive process.

3.2.2. Power Spectral Density PSD Analysis

For Power spectral density PSD analysis, it is necessary to transform the time-domain EN to frequency-domain by applying FFT. There is a correlation with EN signal (with a polynomial filter applied), after which spectral density is calculated with Equations (11) and (12) [66,67].

$$R_{xx}(m) = \frac{1}{N} \sum_{n=0}^{N-m-1} x(n) \cdot x(n+m), \quad \text{when values are from } 0 < m < N \qquad (11)$$

$$\psi_x(k) = \frac{\gamma \cdot t_m}{N} \cdot \sum_{n=1}^{N} (x_n - \overline{x}_n) \cdot e^{\frac{-2\pi k n^2}{N}} \qquad (12)$$

For PSD interpretation, it is necessary to evaluate the slope and the frequency zero limits (ψ^0). The cut frequency indicates the slope begins. The slope could be helpful to find the corrosion mechanism [49]. The slope is defined by β_x and is represented by Equation (13):

$$log\Psi_x = -\beta_x \log f \qquad (13)$$

The frequency zero limits (ψ^0) give material dissolution information because PSD is related to the total energy present in the system [37]. It is essential to clarify that material dissolution is only present in the current PSD [48,65,68]. The following table was proposed by Mansfeld et al. [69] in 1998 to determine the corrosion phenomena occurring on the material surface; this table is adapted to decibels (see Table 6) [70].

Table 6. β intervals to indicate the type of corrosion [70].

Corrosion Type	dB(V)·Decade^{-1}		dB(A)·Decade^{-1}	
	Minimum	Maximum	Minimum	Maximum
Uniform	0	−7	0	−7
Pitting	−20	−25	−7	−14
Passive	−15	−25	−1	1

Figure 6 shows the PSD results; Figure 6a,b presents PSD for EPN and ECN. Samples have values of slope in potential related to uniform corrosion (see Table 7). The ψ^0 values showed that 2OPC-P presents a faster degradation related to a high corrosion kinetic. With samples exposed three months, the behavior is the same that the exposed two months. Also, samples exposed three months showed higher ψ^0 values indicating that the materials will be dissolved faster (see Figure 6c,d).

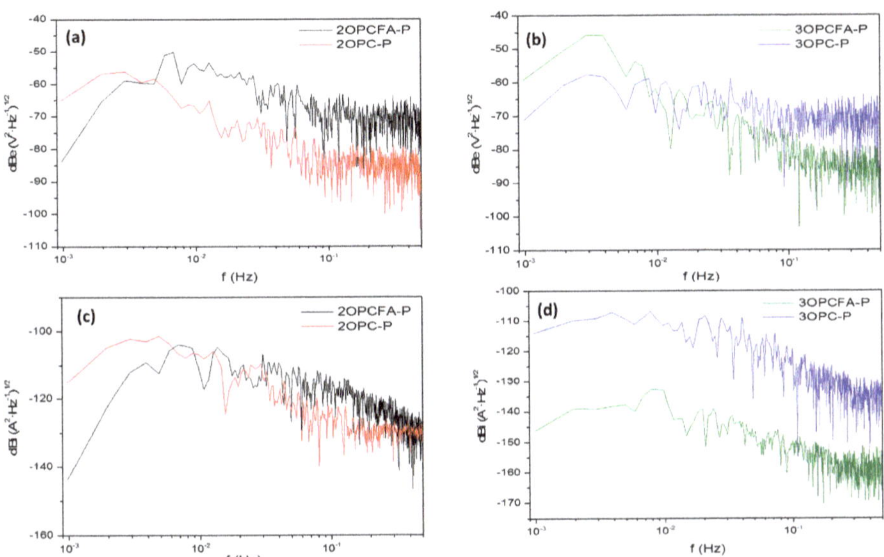

Figure 6. Power spectral density (PSD) for specimens subjected to Impressed current cathodic protection of reinforced concrete mixtures (OPC-P and OPCFA-P). (**a,b**) Potential, 2 and 3 months, (**c,d**) Current, 2 and 3 months.

Table 7. Parameters obtained by PSD for reinforced concrete mixtures (OPC-P/OP and OPCFA-P/OP).

Sample	ψ^0 (dBi)	B (dB [V])
2OPCFA-P	−143.716496	−7
2OPC-P	−64.8956506	−8
3OPCFA-P	−146.383815	−12
3OPC-P	−113.95617	−2
2OPCF-OP	−118.723889	−2
2OPC-OP	−122.621082	−3
3OPCFA-OP	−123.658499	−23
3OPC-OP	−81.2552579	−8

The overprotected samples exposed two months presented slops related to uniform corrosion (Figure 7a,b). 2OPC-OP presents fluctuations at high frequencies even in current or potential; this indicates instability in the processes. The samples exposed for three months present values related to uniform corrosion. Only the sample with FA presented a passivation system's slope value (-23 dB(V)). The material dissolution is higher for OPC samples, and FA mixed showed higher resistance to degradation. Also, overprotected samples present a higher ψ^0 value, indicating higher corrosion kinetic (see Table 7). Values of R_n and ψ^0 presented relation, with a high Rn a low ψ^0 corresponds. Type and corrosion mechanisms presented some uncertainty for slope analysis in PSD [67,70]. For that reason, slope values present limitations to determine the type of corrosion in reinforced concrete mixtures.

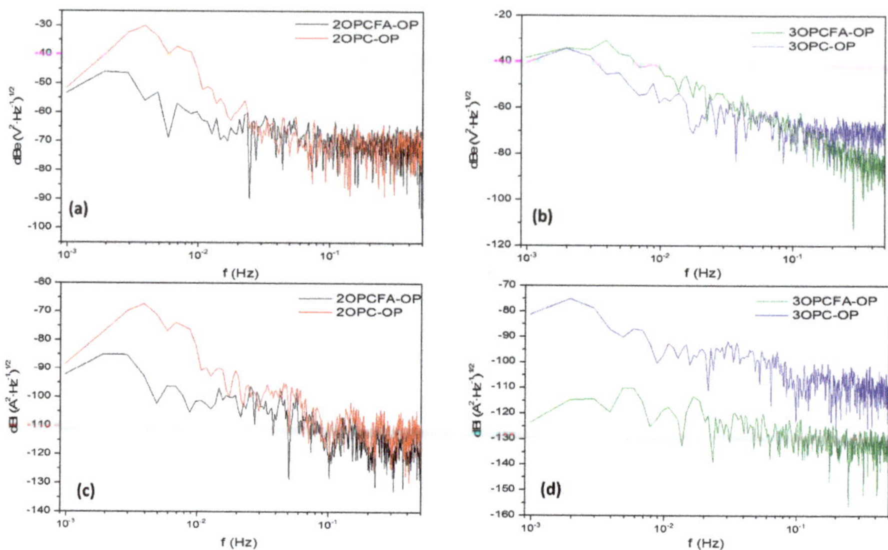

Figure 7. Power spectral density (PSD) for specimens subjected to Impressed current cathodic protection of reinforced concrete mixtures (OPC-OP and OPCFA-OP). (**a**,**b**) ENP, 2 and 3 months, (**c**,**d**) ENC, 2 and 3 months.

4. Discussion

The concrete microstructure for the OPC mixture shows high values in the Ca/Si ratio, being porous concrete. Except for the specimen 2OPCO-P with a ratio of 1.86, the microstructure does not differ in the corrosion process since four specimens indicate uniform corrosion. This behavior indicates that the oxide layer was formed after two months of the steel surface and was maintained both exposure times. Corrosion can be caused because Cl⁻ ion diffuses more quickly through the mixtures microstructure, regardless of protection current and cathodic overprotection. Specimens with FA have a different behavior due to their microstructure and there Ca/Si ratio. FA's Ca/Si ratio shows a better hydrated and less porous concrete, which prevents Cl⁻ ions. Koleva et al. mention that for a high Ca/Si ratio in values of 2.19–2.95, they counteract the positive aspects of a low porosity in the concrete matrix.

The results showed the benefits of employ SCM with OPC. Rubi [71] comments that it is possible to tailor cement with specific properties from the mechanical point of view and durability and chemical stability in highly aggressive environments.

Li and Hou [72,73] study the influence of fly ash in complex cement pastes. The results of Liu and Hou showed that the addition of fly ash means a decrease in the cement content when using cementitious materials, which leads to a decrease in the content of $Ca(OH)_2$ in complex pastes. The cure time of these materials indicates that the crystal of $Ca(OH)_2$ still exists in complex binder pastes at the age of 360 days. With the extension of the curing ages, a large amount of low CaO/SiO_2 ratio of C-S-H gel is generated, which causes complex binder pastes to gradually compact.

Fly ash is a pozzolanic material. It is a finely divided amorphous alumino-silicate with varying amounts of calcium. Calcium mixed with Portland cement and water will react with the calcium hydroxide The hydration of Portland cement releases various calcium-silicate hydrates (C-S-H) and calcium-aluminate hydrates. These pozzolanic reactions are beneficial to the concrete. Pozzolanic reactions increase the quantity of the cementitious binder phase (C-S-H). To a lesser extent, calcium-aluminate hydrates improve the long-term strength and reduce the system's permeability [74–76].

The fly ash is calcium content is perhaps the best indicator of how the fly ash will behave in concrete [77]. However, other compounds such as alkalis (Na_2O and K_2O), carbon, and sulfate (SO_3) can also affect fly ash performance.

Electrochemical noise transients are related to different types of corrosion processes. The pitting process consists of high insensitivity transients with a high repetition rate. If the behavior is mixed, there are transients and oscillations of short amplitude. The uniform process is related to oscillations of low amplitude from the pattern noise. EPN and ECN signals of specimens subjected to cathodic protection current is related to uniform corrosion process show transients low frequency. OPC-P samples presented a higher instability associated with the break of the passive layer. When concrete contains FA and the passive layer is broken, it will be regenerated easier than OPC samples. The EN signal for overprotected samples at 2 and 3 months has current and potential values with low frequencies and amplitudes. Variation of fluctuations is high, associated with an unstable system. Samples with FA presented lower current demand related to lower corrosion kinetic and suggest a passivity phenomenon.

It has been determined that R_n is indeed equivalent to the polarization resistance [43]. Therefore, the $1/R_n$ values are proportional to the corrosion rate according to Ohm's law and the Stern-Geary Equation [61]. The R_n in OPC samples at 2 and 3 months have low values, which indicates a higher corrosion rate. The overprotection FA samples at three months have high Rn values, representing a better behavior in the corrosion kinetics.

In [78], a study about Steel bars embedded in concrete without fly ash and fly ash (the content of 20%) was tested under complete immersion in a 3.5 wt.% NaCl solution. Monitoring of OCP, LRP, and EIS was used to evaluate the steel bars corrosion behavior. The results obtained from electrochemical tests show that partial replacement of fly ash has enhanced corrosion resistance. Also, reduced corrosion rate due to the decrease of permeability to chloride ions. The results obtained from electrochemical noise in OPC and OCPFA samples with cathodic protection complement these investigations to improve the corrosion resistance of steel.

Studies of reinforced concrete applying protection methods focus on adding corrosion inhibitors or rehabilitating structures by electrochemical extraction of chlorine ions or realkalination of concrete. The application of cathodic protection current to reinforced concrete has been commonly studied by electrochemical techniques such as OCP, PP, or EIS. However, EN is not used.

Chloride-induced localized corrosion of embedded steel has been widely known for many years to be one of the most dangerous corrosion types in reinforced concrete [79–81]. Cathodic protection and supplementary materials such as fly ash decrease the effects of corrosion in reinforced concrete produced in harsh environments. Koleva [82–84] investigated the applicability and efficiency of an alternative for ICCP for reinforced concrete based on pulse technology. In specimens made with OPC exposed in 5 wt.% NaCl for a time of 460 days. CP pulse efficiency was evaluated using conventional monitoring techniques (half-cell potential and polarization decay) and electrochemical techniques (potentiodynamic polarization and electrochemical impedance spectroscopy). Recorded parameters even suggest the pulse CP regimens better performance compared to the conventional CP technique.

Using impressed current cathodic protection is essential to define the protection levels to obtain good results and differentiate the effects of cathodic protection and supplementary materials. In this study, the protected and overprotected OPCFA samples have good corrosion resistance. Lu et al. [85] investigated the efficacy of active and passive protection of the the pre-corroded steel reinforcement (1%, 3%, and 6% theoretical mass losses) in reinforced concrete (RC) columns using an externally bonded Carbon Fiber Reinforced Polymer (CFRP) Wrap. Active protection is a novel technique used to prevent corrosion achieved using the CFRP envelope as the anode of the ICCP. Electrochemical measurements were OCP, LPR, and EIS. The results indicate that active protection efficiency is closely related to the pre-corrosion of the protected specimens. Additionally, the efficiency of active protection decreased over time, which can be attributed to the CFRP anode's deterioration and the CFRP/concrete interface. Passive protection was proved to be effective in all cases.

An ICCP system can promote hydrogen activity at the concrete-steel interface. The consumption of electrons from the corrosion process, so safe CP limits, must be established to prevent the risk of phenomena such as hydrogen-embrittlement of steel reinforcements or the concrete-steel interface degradation [27,86–89]. These positional variations in the cathodic current density make it difficult to maintain the same protection level throughout the RCS. In an ICCP system with a low applied current, oxygen and potential levels may be depleted, allowing water reduction to take place:

$$2H_2O + 2e^- \rightarrow H_2 + 2OH^-$$

Moreover, the initiation of fracture can be caused by molecular hydrogen absorbed on the steel surface.

Research with similar additions of fly ash and cathodic protection was carried out by Garcia et al. [27]. Garcia presented results on the adhesion between concrete-steel using the EIS test. Garcia founded that potassium, sodium, and hydrogen ion migration towards the concrete-steel interface play an essential role in bond loss between steel and concrete in structures under cathodic protection impressed current method. Due to a calcium hydroxide reaction, potassium ion content is lower in concrete specimens with fly ash. In contrast, sodium ion migration is higher in concrete containing fly ash due to the sodium ions small size. At the overprotection level, the potassium and sodium ions migrate more quickly than at the protection level. Hydrogen content is higher at the overprotection level than at protection. This phenomenon is attributed to the generation of molecular hydrogen at the overprotection level.

Garces [90] Study of the corrosion level of reinforcing steel bars embedded in Portland cement mortars containing different fly ash types. Concluding the influence of the type of fly ash to establish the durability of the concrete-steel system.

These results demonstrate the using supplementary cementitious materials in OPC increases the corrosion resistance when exposed to chlorides. These results would significantly contribute to creating a more sustainable concrete industry.

5. Conclusions

After studying the Effect of cathodic protection on reinforced concrete with fly ash using electrochemical noise.

- SEM and EDS observations indicated that the microstructures could be observed the hydration products (C-S-H gel) in both mixtures. Hydration influences some results on the Ca/Si ratio obtained, showing that the OPCFA mixtures had better hydration, indicating fewer pores in its microstructure.
- Results indicated that electrochemical noise tests have marked the difference the two mixtures microstructure, obtaining a uniform corrosion process for the specimens elaborated with the OPC mixture. Microstructure facilitated Cl$^-$ ions to the steel surface. Contrary to the mixture made with OPCFA where most of the specimens showed passivity.

- After trend removal, EN signals conserved transients and fluctuation behavior and gave practical corrosion information removing DC.
- EN results show that Rn and Ψ^0 parameters should be considered as a counterpart to calculate corrosion resistance of reinforced concrete mixtures.
- Overprotected samples presented less corrosion resistance than protected samples. This behaviour could be to steel embrittlement by hydrogen adsorption.
- Samples with FA presented a higher trend to passivation than samples with OPC. Besides, FA samples showed higher corrosion resistance.
- Statistical and PSD results showed that uniform corrosion predominates the systems. Although current values are shallow, the variation is considerable, and complicated the analysis for current signals, as potential is more stable than current, makes statistical and PSD analysis easier.
- The current protection and cathodic overprotection did not indicate some tendency in both mixtures behavior, for these exposure times exposure and ratio w/c.
- The discordance of statistical results could be related to developing a different corrosion process on the surface when LI indicates localized corrosion and skewness uniform corrosion. It suggests that localized and uniform corrosion occurs on the surface, but uniform corrosion is the predominant system.

Author Contributions: Conceptualization, F.A.-C., J.G.-C. and C.G.-T.; methodology, I.L.-C., J.C.I.C.; L.D.L.-L., J.O.-C., G.S.-D. and C.G-T.; data curation, F.A.-C., J.J.-M., E.M.-B., D.N.-M., J.O.-C.; formal analysis, F.A.-C., J.G.-C., J.J.-M. and C.G.-T.; writing—review and editing, F.A.-C., J.J.-M. and C.G.-T. All authors have read and agreed to the published version of the manuscript.

Funding: This research was funded by the Mexican National Council for Science and Technology (CONACYT) of the projects CB A1-S-8882 and the Universidad Autónoma de Nuevo León (UANL).

Institutional Review Board Statement: Not applicable.

Informed Consent Statement: Not applicable.

Data Availability Statement: The data presented in this study are available on request from the corresponding author.

Acknowledgments: The authors acknowledge The Academic Body UANL—CA-316 "Deterioration and integrity of composite materials".

Conflicts of Interest: The authors declare no conflict of interest.

References

1. Dacuan, C.N.; Abellana, V.Y. Bond Deterioration of Corroded-Damaged Reinforced Concrete Structures Exposed to Severe Aggressive Marine Environment. *Int. J. Corros.* **2021**, *2021*, 1–13. [CrossRef]
2. Erick Maldonado-Bandala, E.; Higueredo-Moctezuma, N.; Nieves-Mendoza, D.; Gaona-Tiburcio, C.; Zambrano-Robledo, P.O.; Hernández-Martínez, H.; Almeraya-Calderón, F. Corrosion Behavior of AISI 1018 Carbon Steel in Localized Repairs of Mortars with Alkaline Cements and Engineered Cementitious Composites. *Materials* **2020**, *13*, 3327. [CrossRef]
3. Tapan, M.; Aboutaha, R. Effect of steel corrosion and loss of concrete cover on strength of deteriorated RC columns. *Constr. Build. Mater.* **2011**, *25*, 2596–2603. [CrossRef]
4. Azad, A.K.; Ahmad, S.; Azher, S.A. Residual strength of corrosion-damaged reinforced concrete beams. *ACI Mater. J.* **2007**, *104*, 40–47.
5. Tondolo, F. Bond behaviour with reinforcement corrosion. *Constr. Build. Mater.* **2015**, *93*, 926–932. [CrossRef]
6. Bilcik, J.; Holly, I. Effect of reinforcement corrosion on bond behaviou. *Procedia Eng.* **2013**, *65*, 248–253. [CrossRef]
7. Ouglova, A.; Berthaud, Y.; Foct, F.; François, M.; Ragueneau, F.; Petre-Lazar, I. The influence of corrosion on bond properties between concrete and reinforcement in concrete structure. *Mater. Struct.* **2008**, *41*, 969–980. [CrossRef]
8. Zhang, X.; Liang, X.; Wang, Z.; Huang, H.; Zhou, H. An experimental study on effect of steel corrosion on the bondslip performance of reinforced concrete. In Proceedings of the 5th International Conference on Durability of Concrete Structures, Shenzhen University, Shenzhen, China, 30 June–1 July 2016.
9. Mahasenan, N.; Smith, S.; Humphreys, K. The cement industry and global climate change current and potential future cement industry CO_2 emissions. In *Greenhouse Gas Control Technologies, Proceedings of the 6th International Conference, Kyopto, Japan, 1–4 October 2002*; Elsevier: Amsterdam, The Netherlands, 2003; pp. 995–1000.
10. Gartner, E. Industrially interesting approaches to "low-CO_2" cements. *Cem. Concr. Res.* **2004**, *34*, 1489–1498. [CrossRef]

11. Wang, J.; Wu, X.L.; Wang, J.X.; Liu, C.Z.; Lai, Y.M.; Hong, Z.M.; Zheng, J.P. Hydrothermal synthesis and characterization of alkali-activated slag–fly ash–metakaolin cementitious materials. *Microporous Mesoporous Mater.* **2012**, *155*, 186–191. [CrossRef]

12. Baltazar, M.A.; Bastidas, D.M.; Santiago, G.; Mendoza, J.M.; Gaona, C.; Bastidas, J.M.; Almeraya, F. Effect of silica fume and fly ash admixtures on the corrosion behavior of AISI 304 embedded in concrete exposed in 3.5% NaCl solution. *Materials* **2019**, *12*, 4007. [CrossRef]

13. Ariza-Figueroa, H.A.; Bosch, J.; Baltazar-Zamora, M.A.; Croche, R.; Santiago-Hurtado, G.; Landa-Ruiz, L.; Mendoza-Rangel, J.M.; Bastidas, J.M.; Almeraya-Calderón, F.; Bastidas, D.M. Corrosion Behavior of AISI 304 Stainless Steelreinforcements in SCBA-SF Ternary Ecological 3 Concrete exposed to MgSO₄. *Materials* **2020**, *13*, 2412. [CrossRef]

14. Criado, M.; Bastidas, D.M.; Fajardo, S.; Fernández-Jiménez, A.; Bastidas, J.M. Corrosion behaviour of a new low-nickel stainless steel embedded in activated fly ash mortars. *Cem. Concr. Compos.* **2011**, *33*, 644–652. [CrossRef]

15. Troconis de Rincón, O.; Montenegro, J.C.; Vera, R.; Carvajal, A.M.; De Gutiérrez, R.M.; Del Vasto, S.; Saborio, E.; Torres-Acosta, A.; Pérez-Quiroz, J.; Martínez-Madrid, M.; et al. Reinforced Concrete Durability in Marine Environments DURACON Project: Long-Term Exposure. *Corrosion* **2016**, *72*, 824–833. [CrossRef]

16. Santiago, G.; Baltazar, M.A.; Galván, R.; López, L.; Zapata, F.; Zambrano, P.; Gaona, C.; Almeraya, F. Electrochemical Evaluation of Reinforcement Concrete Exposed to Soil Type SP Contaminated with Sulphates. *Int. J. Electrochem. Sci.* **2016**, *11*, 4850–4864. [CrossRef]

17. Talha Junaid, M.; Kayali, O.; Khennane, A.; Black, J. A mix design procedure for low calcium alkali activated fly ash-based concretes. *Constr. Build. Mater.* **2015**, *79*, 301–310. [CrossRef]

18. Habert, G.; d'Espinose de Lacaillerie, J.B.; Roussel, N. An environmental evaluation of geopolymer based concrete production: Reviewing current research trends. *J. Clean. Prod.* **2011**, *19*, 1229–1238. [CrossRef]

19. Maldonado-Bandala, E.; Jiménez-Quero, V.; Olguin-Coca, J.; Lizarraga, L.G.; Baltazar-Zamora, M.A.; Ortiz-C, A.; Almeraya, C.F.; Zambrano, R.P.; Gaona-Tiburcio, C. Electrochemical Characterization of Modified Concretes with Sugar Cane Bagasse Ash. *Int. J. Electrochem. Sci.* **2011**, *6*, 4915–4926.

20. Padhi, R.; Mukharjee, B. Effect of rice husk ash on compressive strength of recycled aggregate concrete. *J. Basic Appl. Eng. Res.* **2017**, *4*, 356–359.

21. Baltazar, M.A.; Mendoza, J.M.; Croche, R.; Gaona, C.; Hernández, C.; López, L.; Olguín, F.; Almeraya, F. Corrosion behavior of galvanized steel embedded in concrete exposed to soil type MH contaminated with chlorides. *Front. Mater.* **2019**, *6*, 6. [CrossRef]

22. Pellegrini Cervantes, J.M.; Barrios Durstewitz, P.C.; Núñez Jaquez, R.; Almeraya Calderón, F.; Rodríguez-Rodríguez, M.; Fajardo-San-Miguel, G.; Martinez-Villafañe, A. Accelerated corrosion test in mortars of plastic consistency with replacement of rice husk ash and nano-SiO₂. *Int. J. Electrochem. Sci.* **2015**, *10*, 8630–8643.

23. Pellegrini-Cervantes, M.J.; Almeraya-Calderon, F.; Borunda-Terrazas, A.; Bautista-Margulis, R.G.; Chacón-Nava, J.G.; Fajardo-San-Miguel, G.; Almaral-Sanchez, J.L.; Barrios-Durstewitz, C.; Martinez-Villafañe, A. Corrosion Resistance, Porosity and Strength of lended Portland Cement Mortar Containing Rice Husk Ash and Nano-SiO₂. *Int. J. Electrochem. Sci.* **2013**, *8*, 10697–10710.

24. Saraswathy, V.; Song, H.W. Corrosion performance of rice husk ash blended concrete. *Constr. Build. Mater.* **2007**, *21*, 1779–1784. [CrossRef]

25. Abu Bakar, B.H.; Putrajaya, R.; Abdulaziz, H. Malaysian rice husk ash–improving the durability and corrosion resistance of concrete: Pre-review. *Concr. Res. Lett.* **2010**, *11*, 6–13.

26. Hossain, M.M.; Karim, M.R.; Hasan, M.; Hossain, M.K.; Zain, M.F.M. Durability of mortar and concrete made up of pozzolans as a partial replacement of cement: A review. *Constr. Build. Mater.* **2016**, *116*, 128–140. [CrossRef]

27. García, J.; Almeraya-Calderon, F.; Barrios, C.; Nuñez, R.; Lopez, M.; Rodríguez, M.; Martínez, A.; Bastidas, J.M. Effect of cathodic protection on steel–concrete bond strength using ion migration measurements. *Cem. Concr. Compos.* **2012**, *4*, 242–247. [CrossRef]

28. Corral, H.R.; Arredondo, R.S.P.; Neri, F.M.; Gómez, S.J.M.; Almeraya, C.F.; Castorena, G.J.H.; Almaral, S.J. Sulfate attack and reinforcement corrosion in concrete with recycled concrete aggregates and supplementary cementing materials. *Int. J. Electrochem. Sci.* **2011**, *6*, 613–621.

29. Corral-Higuera, R.; Arredondo-Rea, P.; Neri-Flores, M.A.; Gómez-Soberón, J.M.; Almaral-Sánchez, J.L.; Castorena-González, J.C.; Almeraya-Calderón, F. Chloride ion penetrability and Corrosion Behavior of Steel in Concrete with Sustainability Characteristics. *Int. J. Electrochem. Sci.* **2011**, *6*, 958–970.

30. Erdoğdu, S.; Kondratovab, L.; Bremner, T. Determination of chloride diffusion coefficient of concrete using open-circuit potential measurements. *Cem. Concr. Res.* **2004**, *34*, 603. [CrossRef]

31. Wyatt, B.S. Cathodic protection of steel in concrete. *Corros. Sci.* **1993**, *35*, 1601–1615. [CrossRef]

32. Qingnan Zhana, E.; Abbas, Z.; Tang, L. Predicting degradation of the anode–concrete interface for impressed current cathodic protection in concrete. *Constr. Build. Mater.* **2018**, *185*, 57–68. [CrossRef]

33. Pedeferri, P. Cathodic protection and cathodic prevention. *Constr. Build. Mater.* **1996**, *10*, 391–402. [CrossRef]

34. Almeraya-Calderón, F.; Estupiñán, F.; Zambrano, R.P.; Martínez-Villafañe, A.; Borunda, T.A.; Colás, O.R.; Gaona-Tiburcio, C. Análisis de los transitorios de ruido electroquímico para aceros inoxidables 316 y -DUPLEX 2205 en NaCl y FeCl. *Rev. Metal.* **2012**, *4*, 147. [CrossRef]

35. Mehdipour, M.; Naderi, R.; Markhali, B.P. Electrochemical study of effect of the concentration of azole derivatives on corrosion behavior of stainless steel in H₂SO₄. *Prog. Org. Coat.* **2014**, *77*, 1761–1767. [CrossRef]

36. Kelly, R.G.; Scully, J.R.; Shoesmith, D.W.; Buchheit, G. *Electrochemical Techniques in Corrosion Science and Engineering*; Taylor & Francis: Boca Raton, FL, USA, 2002; pp. 54–123.

37. Kearns, J.R.; Eden, D.A.; Yaffe, M.R.; Fahey, J.V.; Reichert, D.L.; Silverman, D.C. ASTM Standardization of Electrochemical Noise Measurement. In *Electrochemical Noise Measurement for Corrosion Applications*; Kearns, J.R., Scully, J.R., Roberge, P.R., Reirchert, D.L., Dawson, L., Eds.; ASTM International, Materials Park: Russell, OH, USA, 1996; pp. 446–471.

38. Ma, C.; Song, S.; Gao, Z.; Wang, J.; Hu, W.; Behnamian, Y.; Xia, D.H. Electrochemical noise monitoring of the atmospheric corrosion of steels: Identifying corrosion form using wavelet analysis. *Corros. Eng. Sci. Technol.* **2017**, *5*, 1–9. [CrossRef]

39. Ma, C.; Wang, Z.; Behnamian, Y.; Gao, Z.; Wu, Z.; Qin, Z.; Xia, D.H. Measuring atmospheric corrosion with electrochemical noise: A review of contemporary methods. *Measurement* **2019**, *138*, 54–79. [CrossRef]

40. Xia, D.H.; Song, S.; Behnamian, Y.; Hu, W.; Cheng, F.; Luo, J.L.; Huet, F. Review—Electrochemical Noise Applied in Corrosion Science: Theoretical and Mathematical Models towards Quantitative Analysis. *J. Electrochem. Soc.* **2020**, *167*, 081507. [CrossRef]

41. Botana, P.J.; Bárcena, M.M.; Villero, Á.A. *Ruido Electroquímico: Métodos de Análisis*; Septem Ediciones: Cadiz, Spain, 2002; pp. 50–70.

42. Gaona-Tiburcio, C.; Aguilar, L.M.R.; Zambrano-Robledo, P.; Estupiñán-López, F.; Cabral-Miramontes, J.A.; Nieves-Mendoza, D.; Castillo-González, E.; Almeraya-Calderón, F. Electrochemical Noise Analysis of Nickel Based Superalloys in Acid Solutions. *Int. J. Electrochem. Sci.* **2014**, *9*, 523–533.

43. Montoya-Rangel, M.; de Garza-Montes, O.N.; Gaona-Tiburcio, C.; Colás, R.; Cabral-Miramontes, J.; Nieves-Mendoza, D.; Maldonado-Bandala, E.; Chacón-Nava, J.; Almeraya-Calderón, F. Electrochemical noise measurements of advanced high-strength steels in different solutions. *Metals* **2020**, *10*, 1232. [CrossRef]

44. Monticelli, C. Evaluation of Corrosion Inhibitors by Electrochemical Noise Analysis. *J. Electrochem. Soc.* **1992**, *139*, 706. [CrossRef]

45. Park, C.J.; Kwon, H.S. Electrochemical noise analysis of localized corrosion of duplex stainless steel aged at 475 °C. *Mater. Chem. Phys.* **2005**, *91*, 355–360. [CrossRef]

46. Suresh, G.U.; Kamachi, M.S. Electrochemical Noise Analysis of Pitting Corrosion of Type 304L Stainless Steel. *Corrosion* **2014**, *70*, 283–293. [CrossRef]

47. Cabral-Miramontes, J.A.; Barceinas-Sánchez, J.D.O.; Poblano-Salas, C.A.; Pedraza-Basulto, G.K.; Nieves-Mendoza, D.; Zambrano-Robledo, P.C.; Almeraya-Calderón, F.; Chacón-Nava, J.G. Corrosion Behavior of AISI 409Nb Stainless Steel Manufactured by Powder Metallurgy Exposed in H_2SO_4 and NaCl Solutions. *Int. J. Electrochem. Sci.* **2013**, *8*, 564–577.

48. Nagiub, A.M. Electrochemical Noise Analysis for Different Green Corrosion Inhibitors for Copper Exposed to Chloride Media. *Port. Electrochim. Acta* **2017**, *35*, 201–210. [CrossRef]

49. Dawson, D.L. Electrochemical Noise Measurement: The definitive In-Situ Technique for Corrosion Applications? In *Electrochemical Noise Measurement for Corrosion Applications STP 1277*; Kearns, J.R., Scully, J.R., Roberge, P.R., Reirchert, D.L., Dawson, L., Eds.; ASTM International, Materials Park: Russell, OH, USA, 1996; pp. 3–39.

50. Cottis, R.; Turgoose, S.; Mendoza-Flores, J. The Effects of Solution Resistance on Electrochemical Noise Resistance Measurements: A Theorical Analysis. In *Electrochemical Noise Measurement for Corrosion Applications STP 1277*; Kearns, J.R., Scully, J.R., Roberge, P.R., Reirchert, D.L., Dawson, L., Eds.; ASTM International, Materials Park: Russell, OH, USA, 1996; pp. 93–100.

51. Jáquez-Muñoz, J.M.; Gaona-Tiburcio, C.; Cabral-Miramontes, J.; Nieves-Mendoza, D.; Maldonado-Bandala, E.; Olguín-Coca, J.; López-León, L.D.; Flores-De los Rios, J.P.; Almeraya-Calderón, F. Electrochemical Noise Analysis of the Corrosion of Titanium Alloys in NaCl and H_2SO_4 Solutions. *Metals* **2021**, *11*, 105. [CrossRef]

52. Lara-Banda, M.; Gaona-Tiburcio, C.; Zambrano-Robledo, P.; Delgado-E, M.; Cabral-Miramontes, J.A.; Nieves-Mendoza, D.; Maldonado-Bandala, E.; Estupiñan-López, F.; Chacón-Nava, J.G.; Almeraya-Calderón, F. Alternative to Nitric Acid Passivation of 15-5 and 17-4PH Stainless Steel Using Electrochemical Techniques. *Materials* **2020**, *13*, 2836. [CrossRef] [PubMed]

53. *NMX-C-414-ONNCCE-2014–Industria de la Construcción—Cementantes Hidráulicos - Especificaciones y Métodos de Ensayo*; ONNCCE, Cd.: México City, Mexico, 2014.

54. ASTM. *Standard Specification for Coal Fly Ash and Raw or Calcined Natural Pozzolan for Use in Concrete*; ASTM C618-03 Standard; ASTM International: West Conshohocken, PA, USA, 2008. [CrossRef]

55. ACI. *ACI 211.1-2004 Standard, Standard Practice for Selecting Proportions for Normal, Heavyweight, and Mass Concrete*; American Concrete Institute: Detroit, MI, USA, 2004.

56. ONNCCE. *NMX-C-159-ONNCCE-2004, Industria de la Construcción—Concreto—Elaboración y Curado de Especímenes en el Laboratorio*; ONNCCE, Cd.: México City, Mexico, 2004.

57. ASTM. *Standard Guide for Electrochemical Noise Measurement*; ASTM G199-09; ASTM International: West Conshohocken, PA, USA, 2009.

58. Song, G. Equivalent circuit model for AC electrochemical impedance spectroscopy of concrete. *Cem. Concr. Res.* **2000**, *30*, 1723–1730. [CrossRef]

59. Neville, A. *Properties of the Concrete*; John Wiley & Sons: Hoboken, NJ, USA, 1996.

60. Morgan, J.H. *Cathodic Protection*; National Association of Corrosion Engineers: Houston, TX, USA, 1993.

61. Stern, M.; Geary, A.L. Electrochemical polarization. I. A theoretical analysis of the shape of the polarization curves. *J. Electrochem. Soc.* **1957**, *104*, 56–63. [CrossRef]

62. Eden, D.A.; John, D.G.; Dawson, J.L. Corrosion Monitoring. International Patent WO 87/07022, 19 November 1997. Available online: https://patentimages.storage.googleapis.com/19/ca/d4/c180ce2c0b9dfe/WO1987007022A1.pdf (accessed on 15 March 2021).

63. Mansfeld, F.; Sun, Z. Technical Note: Localization Index Obtained from Electrochemical Noise Analysis. *Corrosion* **1999**, *55*, 915–918. [CrossRef]

64. Reid, S.A.; Eden, D.A. Assessment of Corrosion. US 6,264,824 B1, 24 July 2001. Available online: http://www.khdesign.co.uk/Patents/US6264824.Eden%20AI.pdf (accessed on 15 March 2021).

65. Cottis, R. Interpretation of Electrochemical Noise Data. *Corrosion* **2001**, *57*, 265–285. [CrossRef]

66. Homborg, A.M.; Tinga, T.; Zhang, X.; Van Westing, E.P.M.; Ferrari, G.M.; Wit, J.H.W.; Mol, J.M.W. A Critical Appraisal of the Interpretation of Electrochemical Noise for Corrosion Studies. *Corrosion* **2017**, *70*, 971–987. [CrossRef]

67. Kelekanjeri, V.S.K.G.; Moss, L.K.; Gerhardt, R.A.; Ilavsky, J. Quantification of the coarsening kinetics of γ' precipitates in Waspaloy microstructures with different pri-or homogenizing treatments. *Acta Mater.* **2009**, *57*, 4658. [CrossRef]

68. Legat, A.; Dolecek, V. Corrosion Monitoring System Based on Measurement and Analysis of electrochemical Noise. *Corrosion* **1995**, *51*, 295–300. [CrossRef]

69. Lee, C.C.; Mansfeld, F. Analysis of electrochemical noise data for a passive system in the frequency domain. *Corros. Sci.* **1998**, *40*, 959–962. [CrossRef]

70. Homborg, A.M.; Cottis, R.A.; Mol, J.M.C. An integrated approach in the time, frequency and time-frequency domain for the identification of corrosion using electrochemical noise. *Electrochim. Acta* **2016**, *222*, 627–640. [CrossRef]

71. Mejia de Gutiérrez, R. Effect of supplementary cementing materials on the concrete corrosion control. *Rev. Metal.* **2003**, *39*, 250–255. [CrossRef]

72. Li, X.; Yan, J.; Huaquan Yang, H. Influence of Fly Ash on Microstructure of Complex Binder Pastes. *Adv. Mater. Res.* **2015**, *1088*, 608–612. [CrossRef]

73. Hou, J.; Mei, F.; Mei, Z. Quantitative Research on Strength Characteristics of Phosphogypsum Consolidation Materials. *Appl. Mech. Mater.* **2013**, *307*, 358–363. [CrossRef]

74. Marceau, M.L.; Gajda, J.; VanGeem, M.G. *Use of Fly Ash in Concrete: Normal and High Volume Ranges*; PCA R&D Serial No. 2604; Portland Cement Association: Skokie, IL, USA, 2002.

75. Helmuth, R. *Fly Ash in Cement and Concrete*; SP040; Portland Cement Association: Skokie, IL, USA, 1987; 203p.

76. Malhotra, V.M.; Zhang, M.-H.; Read, P.H.; Ryell, J. Long term mechanical properties and durability characteristics of high strength/high-performance concrete incorporating supplementary cementing materials under outdoor exposure conditions. *ACI Mater. J.* **2002**, *97*, 518–525.

77. Thomas, M.D.A.; Shehata, M.; Shashiprakash, S.G. The Use of Fly Ash in Concrete: Classification by Composition. *Cem. Concr. Aggreg.* **1999**, *12*, 105–110.

78. Yoon-Seok, C.; Jung-Gu, K.; Kwang-Myong, L. Corrosion behavior of steel bar embedded in fly ash concrete. *Corros. Sci.* **2006**, *48*, 1733–1745. [CrossRef]

79. Alvarez, G.M.; Galvele, R.J. The mechanism of pitting of high purity iron in NaCl solutions. *Corros. Sci.* **1984**, *24*, 27–48. [CrossRef]

80. Feliu, V.; Gonzalez, J.A.; Andrade, C.; Feliu, S. Equivalent circuit for modelling the steel-concrete interface. I. experimental evidence and theoretical predictions. *Corros. Sci.* **1998**, *40*, 975–993. [CrossRef]

81. Andrade, C. Calculation of chloride diffusion coefficients in concrete from ionic migration measurements. *Cem. Concr. Res.* **1993**, *23*, 724–742. [CrossRef]

82. Koleva, D.A.; Guo, Z.; Breugel, K.; de Wit, J.H. Conventional and pulse cathodic protection of reinforced concrete: Electrochemical behavior of the steel reinforcement after corrosion and protection. *Mater. Corros.* **2009**, *60*, 344–354. [CrossRef]

83. Koleva, D.A.; Hu, J.; Fraaij, A.; Beek, H. Corrosion and Prevention 2005. In Proceedings of the Conference Australian Corrosion Association, CAP05, Gold Coast, Australia, 20–23 November 2005.

84. Koleva, D.A.; Breugel, K.W.; de Wit, J.H.; Westing, E.; Copuroglu, O.; Fraaij, L.A. Correlation of microstructure, electrical properties and electrochemical phenomena in reinforced mortar. Breakdown to multi-phase interface structures. Part II: Pore network, electrical properties and electrochemical response. *Mater. Charact.* **2008**, *59*, 290. [CrossRef]

85. Lu, Y.; Hu, J.; Li, S.; Tang, W. Active and Passive Protection of Steel Reinforcement in Concrete Column Using Carbon Fibre Reinforced Polymer against Corrosion. *Electrochim. Acta* **2018**, *278*, 124–136. [CrossRef]

86. Zampronio, M.A.; Fassini, F.D.; Miranda, P.E.V. Design of ion-implanted hydrogen contamination barrier layers for steel. *Surf. Coat. Technol.* **1995**, *70*, 203–209. [CrossRef]

87. Chang, J.J. A study of the bond degradation of rebar due to cathodic protection current. *Cem. Concr. Res.* **2002**, *32*, 657–663. [CrossRef]

88. Chang, J.J.; Yeih, W.; Huang, R. Degradation of the bond strength between rebar and concrete due to the impressed cathodic current. *J. Mater. Sci. Tech.* **1999**, *7*, 89–93.

89. Rasheeduzzafar, M.G.A.; Alsulaimani, G.J. Degradation of bond between reinforcing steel and concrete due to cathodic protection current. *ACI Mater. J.* **1993**, *90*, 8–15.

90. Garcés, P.; Andión, L.G.; Zornoza, E.; Bonilla, M.; Payá, J. The effect of processed fly ashes on the durability and the corrosion of steel rebars embedded in cementmodified fly ash mortars. *Cem. Concr. Compos.* **2010**, *32*, 204–210. [CrossRef]

 *materials*

Article

Protection of Concrete Structures: Performance Analysis of Different Commercial Products and Systems

Denny Coffetti [1,2,3], Elena Crotti [1,2,3], Gabriele Gazzaniga [1,3], Roberto Gottardo [4], Tommaso Pastore [1,2,3] and Luigi Coppola [1,2,3,*,†]

[1] Department of Engineering and Applied Sciences, University of Bergamo, 24044 Dalmine (BG), Italy; denny.coffetti@unibg.it (D.C.); elena.crotti@unibg.it (E.C.); gabriele.gazzaniga@guest.unibg.it (G.G.); tommaso.pastore@unibg.it (T.P.)
[2] UdR "Materials and Corrosion", Consorzio INSTM, 50121 Florence, Italy
[3] Consorzio Interuniversitario per lo Sviluppo dei Sistemi a Grande Interfase, CSGI, 50019 Sesto Fiorentino, Italy
[4] Master Builders Solution SPA Italia, 31100 Treviso, Italy; roberto.gottardo@mbcc-group.com
[*] Correspondence: luigi.coppola@unibg.it; Tel.: +39-035-2052054
[†] Current Address: Via Marconi, 5, 24044 Dalmine (BG), Italy.

Abstract: The increasing demand for reconstructions of concrete structures and the wide availability on the market of surface protective products and systems could lead to misunderstandings in the decision of the most effective solution. Surface protectors have become increasingly widespread in recent years in concrete restoration interventions thanks to their properties: they are able to protect the substrate from aggressive agents and consequently extend the useful life of the structures. The aim of this article is first of all to present the surface protective treatments available on the market, outlining their strengths and weaknesses. Subsequently, a characterization of seven different commercial coatings for reinforced-concrete structures is provided, taking into account chemical nature, fields of use and effectiveness, both in terms of physic and elastic performance and resistance to aggressive agents that undermine the durability of the treated concrete elements.

Keywords: coatings; concrete; durability

Citation: Coffetti, D.; Crotti, E.; Gazzaniga, G.; Gottardo, R.; Pastore, T.; Coppola, L. Protection of Concrete Structures: Performance Analysis of Different Commercial Products and Systems. *Materials* **2021**, *14*, 3719. https://doi.org/10.3390/ma14133719

Academic Editor: Angelo Marcello Tarantino

Received: 5 June 2021
Accepted: 27 June 2021
Published: 2 July 2021

Publisher's Note: MDPI stays neutral with regard to jurisdictional claims in published maps and institutional affiliations.

1. Introduction

In recent years, the concept of sustainability has gained ground in the construction sector, and a particular attention has been paid to the durability of the structures [1,2]. In fact, with the same environmental impact deriving from the production, the building materials capable of ensuring a prolonged service life, even if characterized by a strong environmental impact deriving from their production, could be much more sustainable than "green" materials, whose durability is often unknown [3,4]. For this reason, one of the main strategies toward sustainability of building materials is to increase their durability, thus reducing the economic and environmental cost associated with repairing works or, worse, demolition and reconstruction.

The penetration of aggressive external agents such as carbon dioxide or chlorides is what is mainly responsible for the deterioration of reinforced-concrete structures [5–7]. For new buildings, the durability of concrete elements can be guaranteed with proper mix design and construction details; in the case of existing structures with durability deficiency, these problems must be solved in a different way, such as cathodic protection/prevention, electrochemical-based techniques, migrant corrosion inhibitors and surface treatments. Cathodic protection/prevention is generally applied only to infrastructures; it requires periodic inspection and proper design to avoid insufficient protection or overprotection and related adverse side effects such as hydrogen embrittlement and stress corrosion cracking [8,9]. So far, electrochemical methods used for concrete re-alkalinization or chloride

following are the main properties of the surface treatments commercially available for concrete protection.

1.2.1. Physical Parameters

First of all, it must be stated that the mechanical performance of the substrate cannot be improved in relation to the performance of the treatment, since the latter is not able to modify the porosity of the concrete; however, it is able to prevent the deterioration of concrete performance. In this regard, the compressive strength is not a noteworthy parameter in the examination of the treatments.

A more important parameter for the surface treatments is the bonding strength; good adhesion to the substrate ensures its long-term performance and protection of concrete [33]. Experimental investigation [34] showed that the average bond strength of different epoxy coatings ranges between 2.9 and 4.0 MPa, and for cementitious mortar, it is about 0.8–1.2 MPa. The adhesion could be influenced by different aspects, such as the utilization of a primer, the roughness and the quality of the substrate and the application technique [35]. It is known that the adhesion property is strongly conditioned by aging of the treatment; unfortunately, no significant number of studies have been conducted in order to understand how much age influences the adhesion property and what the problems related to the aging of the same are [36,37].

Another important indicator of the effectiveness of a treatment is the abrasion resistance, since it is able to evaluate of the service life of surface-treated concrete under repetitive traffic loadings [38,39]. Many of surface treatments can improve the wear resistance of concrete surface. It is important to understand the mechanism according to which the concrete is protected from abrasion since it varies by the treatment type. Dang et al. [40] found that most organic surface coatings could improve the abrasion resistance of concrete, especially epoxy, while methacrylate with a high molecular weight showed no protection. A slight enhancement of abrasion resistance was observed for the concrete treated with silanes. Franzoni et al. [41] investigated the effects of some inorganic surface treatments on abrasion resistance and found out that sodium silicate showed the best effect because it could form a protective layer with a remarkable thickness.

The crack-bridging ability is the property of the coatings to cover cracks formed in the concrete substrate [42], keeping the properties intact and reducing the risk that cracks can propagate and cause deterioration. The crack-bridging ability is closely related to the type of protective material chosen and its elasticity characteristics: the polymer-cement coatings have an excellent crack bridging ability, making this type of treatment suitable for application on cracked supports; the reduced elasticity of epoxy resins and acrylic coatings makes ineffective the crack-bridging properties, reducing the range of applications of the latter to sound substrates only [43–45].

The resistance of the treatment against possible alterations promoted by the environment in which they are located or by the tasks they must perform is very important for the success of a protection intervention. Research showed that temperature and ultraviolet radiation highly affected the efficiency of surface treatments. Vries [46] studied resistance of hydrophobic surface treatments, such as silane and siloxane, at the high temperature. The results showed that the water absorption rate for the treated concrete increased dramatically after half-an-hour storage in a 160 °C chamber. Levi et al. [47] found that protection of silane, silicone and fluorinated polymer on concrete water absorption decreased by 50% after ultraviolet aging; by contrast, the polymer-modified cementitious coatings have a great ultraviolet light resistance, which allows for their use in direct contact with sunlight.

1.2.2. Durability Parameters

In most cases, the success of the protection is related to the durability of the concrete-surface treatment systems [48,49]. Since many aggressive substances are transported through water or air, the permeation characteristics of the surface concrete is an important factor for the durability of whole concrete element [50]. Many surface coatings are able to reduce the ingress of water through the treated matrix. The best treatments from the point of view of water permeability are epoxy coatings, silane with an acrylic top coat, methyl methacrylate, and alkyl alkoxysilane [51]. Almusallam et al. [52] found that the uncoated cement mortars absorbed water at a very rapid rate, and after 56 h, the total absorption was about 5% by weight; after being treated with polymer emulsion, acrylic, chlorinated rubber, polyurethane coatings, and epoxy coatings, the absorption of water decreased to 3.3–3.4%, 0.23–1.46%, 0.76–1.04%, 0.21–1.83%, and 0.27–1.3%, respectively. Moreover, Medeiros et al. [53,54] demonstrated that silane and siloxane have a good capability of inhibiting water penetration as long as the water pressure was lower than 12,000 Pa. The results indicated that hydrophobic surface treatments should only be used when the water exposure conditions are well known. Modified cementitious mortar coating, on the other hand, have slightly higher resistance to water penetration than the other treatments; despite this, the modified cementitious mortar coatings are highly recommended in the protection of structures in everlasting contact with water because, unlike other treatments, they have a greater resistance to leaching [55,56].

In a concrete structure, chlorides can penetrate into the cement matrix and initiate degradation phenomena due to diffusion under the influence of a concentration gradient, migration in an electrical field or absorption due to a capillary action [57]. In most cases, the protective treatments are able to reduce the concentration of chlorides in the substrate. Unfortunately, both for the great variety of treatments and test methods, it is difficult to identify the best treatment even if polymer coatings seem to exhibit the higher protection against chloride ingress in the cementitious matrix. Almusallam et al. affirmed that the polyurethane and acrylic coatings are able to increase the resistance against the diffusion of chloride ions of about 10 times compared to the uncoated concrete [52]. Brenna et al. [58] found the good behavior of polymer modifies cementitious coatings also after 17 years of exposure, highlighting also the influence of the polymer content on the properties of coatings. According to Coppola et al. [12], the silane-based treatment is able to significantly reduce the penetration of chlorides into the matrix, regardless of the type of cement used.

Especially in urban or industrialized areas, the problem of carbon dioxide penetration is relevant. Carbonation is a chemical reaction between $Ca(OH)_2$, calcium-silicate-hydrate (C-S-H) and CO_2 to form $CaCO_3$, silica-rich C-S-H and amorphous silica gel [59]; it is able to destroy the passivity of the embedded reinforcement bars and to promote corrosive phenomena [60,61]. The factors controlling carbonation are the diffusivity of CO_2 and the reactivity of CO_2 with the concrete. They depend on the pore system of hardened concrete and the exposure condition [62]. Many studies agree that acrylic coatings are the best choice to prevent the penetration of carbon dioxide, while treatments based on silanes or siloxanes can control the humidity of the concrete substrate but are not able to reduce the permeability to air and consequently not even the penetration of carbon dioxide [63–65].

Although the treatments cannot fulfil the functions performed by the air-entraining additives against the phenomena related to freezing and thawing cycles, they can contribute to making the concrete substrate more resistant to cold climate by preventing the critical moisture content from being reached. Both the acrylic treatments and the polymer-modified cementitious coatings have a good behavior against the destructive phenomena induced by the freezing and thawing cycles; by contrast, the behavior of silane-based coatings has not yet been well defined and is debated by the scientific community: Basheer et al. [66] stated that the silane treatment could double the number of freeze–thaw cycles at which concrete began cracking in fresh water test; other studies show that silane-treated concrete deteriorates more quickly than untreated concrete in laboratory accelerated freeze-thaw tests. This type of treatment would be very effective if the concrete substrate would be completely dry since they would not allow water to penetrate; however, in reality the substrate is never completely dry and therefore the silane-based treatment is not able to reduce the risk associated with freezing and thawing curls [67,68].

2. Experimental Part

Materials and Methods

In order to provide a complete analysis able of evaluating some of the aspects previously described, 7 different commercial treatments (Table 2) have been selected and tested in order to outline a guideline for selecting the best surface treatment in accordance with the expected purposes. The products investigated are: a water-based acrylic protective (A), a water-based acrylic elastomeric protective (AE), an epoxy coating (E), an epoxy-bituminous coating (EB), an epoxy-polyurethane coating (EP), a polyurethane coating (P) and a polymer-modified cementitious coating (PMC). The treatments were applied on a concrete substrate manufactured in accordance with EN 1766 [69] (CEM I 42.5 R, water-to-cement ratio 0.40, natural aggregates with maximum size 10 mm and superplasticizer in compliance with EN 934-2 [70]). The summary of the experimental tests carried out is reported in Table 3.

Table 2. Properties of surface treatments.

Treatment	Density	Solids	Thickness (t)
Water-based acrylic (A)	$1.54 \, kg/dm^3$	74%	200 µm
Water-based acrylic elastomeric (AE)	$1.35 \, kg/dm^3$	63%	200 µm
Epoxy coating (E)	$1.50 \, kg/dm^3$	84%	400 µm
Epoxy-bituminous (EB)	$1.00 \, kg/dm^3$	42%	350 µm
Epoxy-polyurethane (EP)	$1.30 \, kg/dm^3$	80%	300 µm
Polyurethane (P)	$1.30 \, kg/dm^3$	58%	400 µm
Polymer-modified cementitious (PMC)	$2.05 \, kg/dm^3$	75%	2 mm

Table 3. Details on the experimental tests carried out on coatings.

Test	Standard	A	AE	E	EB	EP	P	PMC
Pot life	EN ISO 9514 [71]	x	x	x	x	x	x	x
Adhesion to concrete	EN 1542 [72]	x	x	x	x	x	x	x
Adhesion to concrete after freeze/thaw cycles with deicing salts	EN 13687-1 [73]	x	x	x	x	x	x	x
CO_2 permeability	EN 1062-6 [74]	x	x	x	x	x	x	x
Water vapor permeability	EN ISO 7783 [75]	x	x	x	x	x	x	x
Capillary water absorption	EN 1062-3 [76]	x	x	x	x	x	x	x
Crack-bridging ability (static and dynamic)	EN 1062-7 [77]	x	x	x	x	x	x	x
Abrasion resistance	EN ISO 5470-1 [78]		x	x	x	x		x
Resistance to severe chemical attack	EN 13529 [79]			x		x		x
Resistance to negative hydraulic pressure	UNI 8298-8 [80]			x		x		x
Fire resistance	EN 13501-1 [81] and related standards	x	x	x	x	x	x	x
UV light and moisture resistance	EN 1062-11 par. 4.2 [82]	x	x	x	x	x	x	x

3. Results

3.1. Pot Life

The pot life is defined as the maximum time during which a coating material supplied as separate components (product A and product B or powder and water) should be used after the components have been mixed together. This parameter was evaluated at room temperature by taking into account the longest period of time which the consistency met the requirements related to the laying of the coating. Results reported in Table 4 evidenced that the average pot life at room temperature of the surface treatments is close to 1 h with the exception of P, which is applicable for about 3 h after mixing. However, it is necessary to point out that the pot life is strongly affected by the job-site temperature: the higher the temperature, the shorter the pot life.

Table 4. Pot life of coatings.

Properties	A	AE	E	EB	EP	P	PMC
Pot life [min]	60	60	45	90	60	180	60

3.2. Adhesion

The adhesion tests were carried out on samples before and after thermal cycles in presence of deicing salts. In particular, after the curing of concrete and the application of the protective coating, a core drilling is carried out that affects both the substrate and the protective one. Then, a plug is applied, and finally, at a speed of 0.05 MPa/s, the tear test is performed on one half of samples to evaluate the adhesion and the failure type. On the other hand, the other specimens were subjected to 50 thermal cycles (immersion in a saturated sodium chloride solution at a temperature of $-15 \pm 2\,°C$ for 2 h, followed by immersion in water at a temperature of $21 \pm 2\,°C$ for 2 h) before the adhesion test.

Figure 1 shows the adhesion strength of the tested treatments in both the aforementioned cases. According to Garbacz [33], the results belong to the studied range; moreover, it is possible to observe that the EP and E coatings have the best adhesion both before and after the thermal cycles with failure on concrete substrate, acrylic and acrylic elastomeric products are characterized by adhesion close to 3 MPa with failure on the coating/concrete interface and EB, P and PMC show similar adhesion values with different failure mechanism: the failure of epoxy-bitumen coating was on the coating/concrete interface, while the polyurethane and polymer-modified cementitious coatings exhibited a cohesive failure.

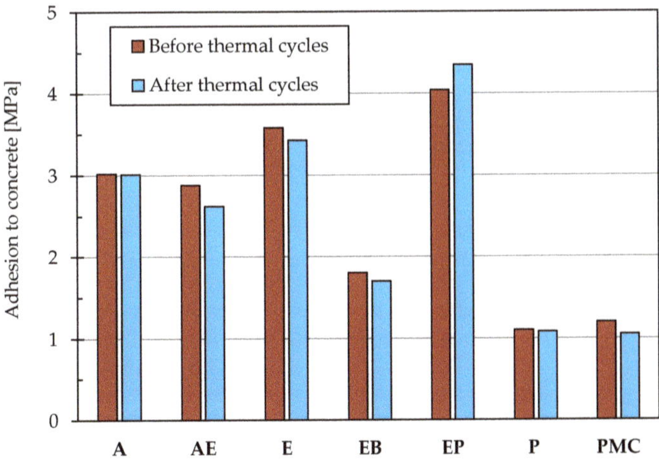

Figure 1. Adhesion to concrete before and after 50 freeze/thaw cycles with deicing salts.

3.3. CO_2 and Water Vapor Permeability

The resistance to carbon dioxide diffusion is an important parameter in the choice of a coating. To avoid any design errors or the impossibility of carrying out the restoration work in a workman-like manner, it is possible to integrate the missing concrete cover thickness by choosing a suitable protective. In light of this, EN 1504-2 [17] sets a minimum value of resistance to carbon dioxide equal to 50 m of equivalent air thickness; most of the coatings on the market, to operate in favor of safety, have values that are much higher than those required by the standard for a coating. In addition to the diffusion of carbon dioxide, it is necessary to prescribe a minimum value for the resistance offered to vapor diffusion. This need arises from the fact that if on the back of the coating the cement matrix is saturated or partially saturated with water, when this reaches the concrete/coating interface, it could be transformed into water vapor due to solar radiation and thus determining the swelling or the detachment of the vapor-tight coat. To avoid these problems, EN 1504-2 [17] requires for vapor diffusion values to be less than 5 m of equivalent air thickness.

The results, first of all, highlight that all the protectives tested comply with the limits prescribed by EN 1504-2 [17] for the protection of reinforced-concrete structures. In particular, Figure 2 evidenced that E, EP and P are impervious to carbon dioxide (Sd CO_2 greater than 300 m), epoxy-bitumen and acrylic elastomeric-based treatments have a resistance to CO_2 penetration of just over 50 m. At the same time, in this study, the E and PMC are characterized by an Sd CO_2 close to 100 m of equivalent air thickness. On the other hand, authors want to highlight that unpublished experimental results on polymer-modified cementitious coatings demonstrated that the Sd CO_2 is strongly influenced by the type and dosage of polymer used in manufacturing the coating. In particular, it is possible to obtain carbon dioxide resistances up to those of polyurethane-based coatings.

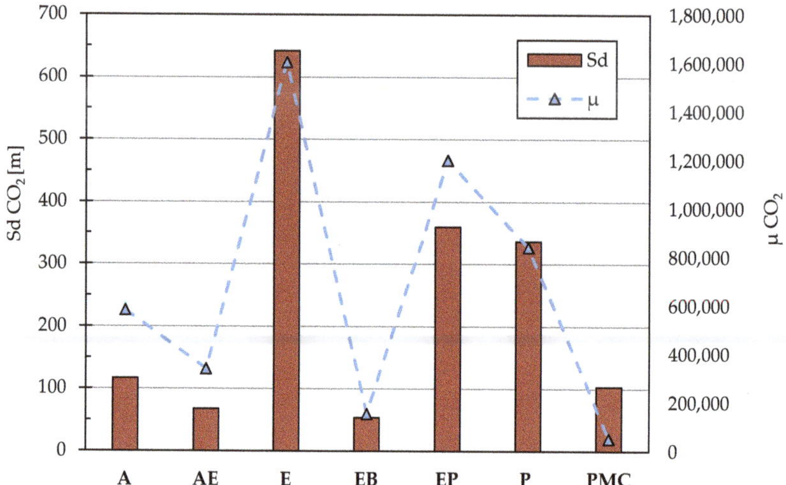

Figure 2. Resistance against CO_2 permeability of coatings expressed in terms of Sd CO_2 (left) and μ CO_2 (right). Sd CO_2 is calculated as the product of μ CO_2 and the application thickness (t).

In terms of water vapor permeability (Figure 3), the best performance is offered by the polymer-modified cementitious coating that has the lowest values together with acrylic coatings. To the detriment of other excellent performances, epoxy coatings can suffer from low vapor permeability which strongly increase, in the presence of wet concretes and/or exposure to solar radiation, the risk of detachment and blistering of the coating [83,84].

3.4. Capillary Absorption

Aggressive substances for the cement matrix are often transported by water by capillary absorption. For this reason, the EN 1504-2 [17] standard limits the water capillary absorption of coated concrete by setting the maximum capillary absorption of coating to 0.1 kg/m^{2}h$^{0.5}$. Results reported in Table 5 show that all the products investigated meet the standard requirements. In particular, epoxy-based coats (E, EB and EP), similarly to polyurethane products, are characterized by a negligible water absorption (0.001–0.008 kg/m^{2}h$^{0.5}$), while A and AE treatments evidence a capillary absorption coefficient close to the standard limits of 0.1 kg/m^{2}h$^{0.5}$. By contrast, the waterproofing ability of polymer-modified cementitious coatings are lower with respect to the epoxy-based treatments, even if it is necessary to demonstrate that this property is also strongly affected by the type and dosage of polymer used in manufacturing the coating. In particular, a proper composition allows one to reach extremely low capillary absorption coefficients.

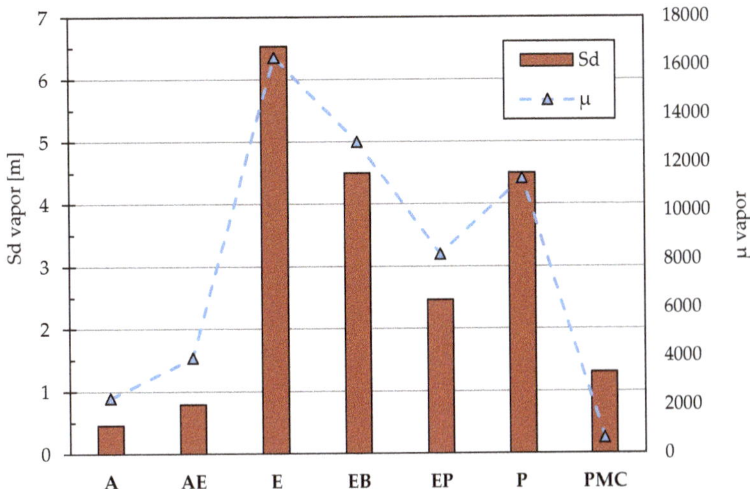

Figure 3. Resistance against water vapor permeability of coatings expressed in terms of Sd vap (left) and μ vap (right). Sd CO_2 is calculated as the product of μ vap and the application thickness (t).

Table 5. Capillary absorption coefficient and results of abrasion test.

Surface Treatment	Capillary Absorption Coefficient [kg/m^{2}h$^{0.5}$]	Mass Loss at the End of Abrasion Test [g]
A	0.090	–
AE	0.090	–
E	0.001	298
EB	0.008	2874
EP	0.006	238
P	0.003	254
PMC	0.010	2949

3.5. Crack-Bridging Ability

The ability to cover the cracks that occur in the substrate to avoid the penetration of aggressive agents turns out to be a fundamental parameter in the choice of a protective. The graph reported in Figure 4 highlights the static crack-bridging values at a temperature of 23 °C in compliance with EN 1062-7 [77]. The AE and P coatings show an excellent ability to cover cracks wider than 1.6 mm; PMC crack-bridging ability is close to 1300 μm, while EB limits their static crack bridging at about 1 mm. By contrast, the intrinsic lack of elasticity of epoxy and acrylic systems is evident in this test. Finally, the epoxy-polyurethane coating is only able to bridge cracks with a width lower than 600 μm.

In addition to the static crack-bridging ability, the capability of the coatings to take up the elongation resulting from the periodic movement of the crack sides was evaluated for different crack widths. After 1000 cycles with a maximum crack width of 500 μm, AE and P appear to be sound, while the A and PMC systems limit their dynamic ability to cover cracks of about 150 μm (Figure 4). On the other hand, the epoxy nature of the other systems (E, EP, EB) does not guarantee adequate dynamic crack-bridging ability.

3.6. Abrasion Resistance

In relation to the weight loss connected to a forced abrasion (EN 5470-1 [78]) listed in Table 5, it is possible to note that epoxy, polyurethane and epoxy-polyurethane protective products are more resistant to abrasion (mass loss in the range of 200–300 g at the end of the test), while PMC and EB coatings show a weight loss that is about one order of magnitude higher than the aforementioned systems. The reason for this strong damage following an abrasive tickling is to be found in the physical properties of the coatings; in fact, as previously highlighted by the capillary absorption tests, PMC and EB are more porous, therefore less compact and consequently more exposed to these stresses [40,41].

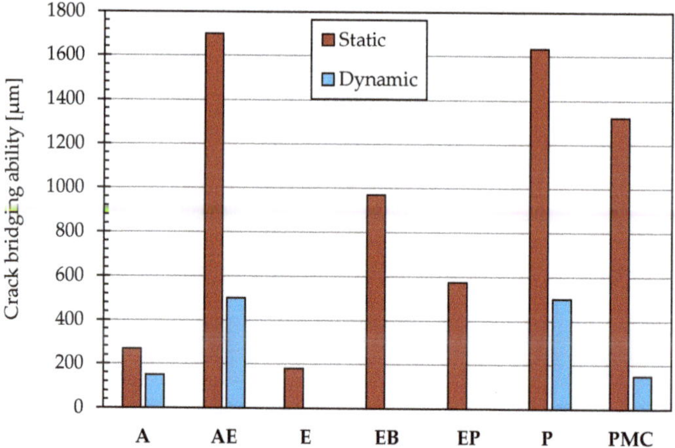

Figure 4. Crack-bridging test results.

3.7. Resistance to Severe Chemical Attack

The evaluation of the resistance to make contact with chemical agents is an important parameter closely related to the intended applications of the coating; due to these peculiarities, some protective products have not been tested because their low resistance against chemicals is commonly known [85]. Table 6 shows the hardness values (EN ISO 868 [86]) after the exposure to different aggressive chemical agents in accordance with UNI 13529 [79]. Results are expressed from 0 to 100: a high shore hardness means a great resistance to chemicals exposure. The treatment that shows greater protection against aggressive chemicals is the EP one, since, following exposure to chemicals, it reports hardness values that are on average 5% higher than the other coatings analyzed. E and MPC coatings also guarantee good chemical resistance [87,88], even if it should be noted that cement polymer-modified coatings show significant deterioration when exposed to acidic environments [89]. In fact, following an exposure to 10% aqueous acetic acid and 20% sulfuric acid, the total dissolution of the coating layer takes place.

Table 6. Resistance to chemical aggression (hardness shore D).

Chemical	E			EP			PMC		
	0 Day	3 Days	28 Days	0 Day	3 Days	28 Days	0 Day	3 Days	28 Days
All hydrocarbons	73	73	77	78	85	84	76	80	75
All alcohols and glycol ethers	74	77	75	75	80	79	–	–	–
Hydrogenated hydrocarbons	74	76	76	74	N.D.	N.D.	–	–	–
Aqueous solutions of organic acids up to 10%	76	74	66	81	79	69	78	N.D.	N.D.
Inorganic acids up to 20% and acidic hydrolyzing salts in aqueous solutions	76	78	75	83	82	76	78	N.D.	N.D.
Inorganic bases and the alkaline hydrolyzing salts in aqueous solutions	73	78	77	80	79	82	78	76	77
Solutions of inorganic nonoxidizing salts	76	77	77	81	84	78	82	81	82
water at the inlet of the purifier	75	76	77	81	84	78	76	77	76

N.D. = Not detectable due to severe deterioration.

3.8. Resistance to Hydraulic Pressure

The analysis of the parameter of resistance to negative hydrostatic pressure becomes of primary importance for waterproofing underground structures such as retaining walls. For this purpose, not all the tested coatings are suitable; therefore, the test in compliance with EN 8298-8 [80] was conducted only on E, EP and MPC coatings. These coatings, following the application of different hydrostatic pressures (2, 5, 10, 50, 100, and 250 kPa for 72 h), did not show any state of alteration connected to the permeation of pressurized water, making them suitable for this purpose.

3.9. Fire Resistance

The fire resistance is a fundamental properties of building materials, and it was evaluated by determining the ignitability of products by direct flame impingement and the production of smoke and flaming droplets. Experimental results reported in Table 7 evidenced a limited resistance to fire of PMC and EB, because they contain huge number of inflammable polymers (PMC) or oil-derived products (EB). On the other hand, acrylic elastomeric, epoxy, epoxy-polyurethane and polyurethane coatings showed a moderate fire resistance while acrylic product is characterized by a high resistance to flames. Moreover, all the specimens (except EB and PMC) are characterized by a low smoke production (class s1) and no formation of flaming droplets and particles (class d0). However, it is necessary to outline that the fire resistance of polymer-modified cementitious coatings is strongly influenced by the type and the dosage of the polymer used; in general, the fire resistance class of PMC is in the range F–C (class A1 indicated the maximum fire resistance, class F the minimum).

Table 7. Fire resistance of coatings (main results).

Surface Treatment	Classification	Flame Spread	Direct Flame Test Results		Flaming Droplets
			FIGRA *	SMOGRA **	
A [#]	A2 s1 d0	≤150 mm within 60 s	≤120 W/s	≤30 m^2/s^2	No droplets
AE	B s1 d0	≤150 mm within 60 s	≤120 W/s	≤30 m^2/s^2	No droplets
E	C s1 d0	≤150 mm within 60 s	≤250 W/s	≤30 m^2/s^2	No droplets
EB	E	≤150 mm within 20 s	–	–	–
EP	B s1 d0	≤150 mm within 60 s	≤120 W/s	≤30 m^2/s^2	No droplets
P	B s1 d0	≤150 mm within 60 s	≤120 W/s	≤30 m^2/s^2	No droplets
PMC	E	≤150 mm within 20 s	–	–	–

* FIGRA (fire growth rate index) is the maximum of the quotient of heat release rate from the sample and the time of its occurrence using a total heat release threshold in the first 600 s of 0.4 MJ (for class C) or 0.2 MJ (for class B and A2); ** SMOGRA (smoke growth rate index) is the maximum of the quotient of smoke production rate from the sample and the time of its occurrence; # every class A2 product shall satisfy the same criteria as for class B with the addition of non-combustibility tests in accordance with EN ISO 1182 [90].

3.10. Resistance to UV Light and Moisture

All the samples were exposed to wetting and drying cycles and UV radiation in accordance with EN 1062-11 [82] to determine the resistance of coatings to UV light and moisture. After more than 1000 h of exposure (about 125 wetting and drying cycles), no cracks, detachments and flaking were found.

4. Conclusions

In light of the scientific literature review and tests performed on protective products, the following conclusions can be drawn:

(a) Water-based acrylic and water-based acrylic elastomeric coatings are the best systems when it is necessary to maintain the original texture and/or a specific aesthetic finish is desired. Acrylics have an excellent resistance to CO_2 penetration and a proper water absorption; they usually do not have excellent crack-bridging capabilities and, due to their conformation, are not suitable for withstanding mechanical stress or contact with aggressive chemical agents. They can be used to protect concrete in structures not fully immersed in water or in contact with aggressive chemicals; finally, elastomeric acrylic coatings embrace a slightly wider market share given the crack-bridging properties that make them suitable for applications where an elasticity of the coating is required, such as the protection of residential buildings.

(b)   Polyurethane and epoxy resin-based coatings, whether they are epoxy, epoxy-polyurethane or epoxy-bituminous can be used to protect concrete structures subject to continuous contact with water and aggressive chemicals. The main applications of this family of coatings are the protection a of reinforced-concrete structures subjected to severe environmental aggressions such as sewage collectors, purification plants (settling and aeration tanks), and bridge decks. The limits of epoxy protectives are the high resistance to vapor diffusion which could compromise the durability of coatings applied on wet substrates and the reduced crack-bridging capability that preclude their use on cracked concretes. The EB, EP and P, on the contrary, shows good crack-bridging ability, which makes them suitable for applications where a good ability to cover cracks is required, such as waterproofing of canals or containment tanks.

(c)   Polymer-modified cementitious coatings have excellent resistance to steam penetration and crack-bridging ability, as well as a high resistance to hydraulic pressure. They can be applied in the protection of both new concrete structures or during restoration works. The high waterproofing properties make these products suitable for the protection of reinforced-concrete structures such as tanks, fountains, wells, flower boxes or in any case of structures in continuous contact with water without extremely aggressive chemical agents. Their elasticity makes these protective products suitable for under-tile applications typical of the restoration works of projecting structures such as terraces and balconies. Therefore, it is necessary to highlight that the main properties of these coatings are strongly related to the type and dosage of polymer used during their manufacturing.

Author Contributions: Conceptualization, L.C.; formal analysis, R.G.; investigation, G.G., D.C., and E.C.; writing—original draft preparation, G.G.; writing—review and editing, D.C.; supervision, L.C. and T.P. All authors have read and agreed to the published version of the manuscript.

Funding: This research received no external funding.

Data Availability Statement: Not applicable.

References

1.   Chen, C.; Habert, G.; Bouzidi, Y.; Jullien, A. Environmental impact of cement production: Detail of the different processes and cement plant variability evaluation. *J. Clean. Prod.* **2010**. [CrossRef]

2.   Damtoft, J.S.; Lukasik, J.; Herfort, D.; Sorrentino, D.; Gartner, E.M. Sustainable development and climate change initiatives. *Cem. Concr. Res.* **2008**, *38*, 115–127. [CrossRef]

3.   Coppola, L.; Coffetti, D.; Crotti, E.; Gazzaniga, G.; Pastore, T. An Empathetic Added Sustainability Index (EASI) for cementitious based construction materials. *J. Clean. Prod.* **2019**, *220*, 475–482. [CrossRef]

4.   Gettu, R.; Pillai, R.G.; Meena, J.; Basavaraj, A.S.; Santhanam, M.; Dhanya, B.S. Considerations of Sustainability in the Mixture Proportioning of Concrete for Strength and Durability. *Spec. Publ.* **2018**, *326*, 5.1–5.10.

5.   Bertolini, L.; Elsener, B.; Redaelli, E.; Polder, R. *Corrosion of Steel in Concrete: Prevention, Diagnosis, Repair*, 2nd ed.; Wiley-VCH Verlag GmbH & Co.: Weinheim, Germany, 2013.

6.   Zacchei, E.; Nogueira, C.G. Chloride diffusion assessment in RC structures considering the stress-strain state effects and crack width influences. *Constr. Build. Mater.* **2019**, *201*, 100–109. [CrossRef]

7.   Wu, L.; Wang, Y.; Wang, Y.; Ju, X.; Li, Q. Modelling of two-dimensional chloride diffusion concentrations considering the heterogeneity of concrete materials. *Constr. Build. Mater.* **2020**, *243*, 118213. [CrossRef]

8.   Angst, U.M. A Critical Review of the Science and Engineering of Cathodic Protection of Steel in Soil and Concrete. *Corrosion* **2019**, *75*, 1420–1433. [CrossRef]

9.   Redaelli, E.; Bertolini, L. Electrochemical repair techniques in carbonated concrete. Part II: Cathodic protection. *J. Appl. Electrochem.* **2011**, *41*, 829–837. [CrossRef]

10. Redaelli, E.; Bertolini, L. Electrochemical repair techniques in carbonated concrete. Part I: Electrochemical realkalisation. *J. Appl. Electrochem.* **2011**, *41*, 817–827. [CrossRef]
11. Xia, J.; Cheng, X.; Liu, Q.; Xie, H.; Zhong, X.; Jin, S.; Mao, J.; Jin, W. Effect of the stirrup on the transport of chloride ions during electrochemical chloride removal in concrete structures. *Constr. Build. Mater.* **2020**, *250*, 118898. [CrossRef]
12. Coppola, L.; Coffetti, D.; Crotti, E.; Gazzaniga, G.; Pastore, T. Chloride Diffusion in Concrete Protected with a Silane-Based Corrosion Inhibitor. *Materials* **2020**, *13*, 2001. [CrossRef] [PubMed]
13. Tittarelli, F.; Mobili, A.; Bellezze, T. The use of a Phosphate-Based Migrating Corrosion Inhibitor to Repair Reinforced Concrete Elements Contaminated by Chlorides. *IOP Conf. Ser. Mater. Sci. Eng.* **2017**, *225*, 12106. [CrossRef]
14. Pan, X.; Shi, Z.; Shi, C.; Ling, T.C.; Li, N. A review on concrete surface treatment Part I: Types and mechanisms. *Constr. Build. Mater.* **2017**, *132*, 578–590. [CrossRef]
15. Pan, X.; Shi, Z.; Shi, C.; Ling, T.C.; Li, N. A review on surface treatment for concrete—Part 2: Performance. *Constr. Build. Mater.* **2017**, *133*, 81–90. [CrossRef]
16. Malheiro, R.; Meira, G.R.; Lima, M. Experimental stydy of chloride transport into double layered components with concrete and rendering mortar. *Rev. Constr.* **2016**, *15*, 119–132. [CrossRef]
17. *EN 1504-2-Products and Systems for the Protection and Repair of Concrete Structures—Definitions, Requirements, Quality Control and Evaluation of Conformity-Part 2: Surface Protection Systems for Concrete*; ISO: Geneva, Switzerland, 2005.
18. Yamini, S.; Young, R.J. Stability of crack propagation in epoxy resins. *Polymer* **1977**, *18*, 1075–1080. [CrossRef]
19. Moloney, A.C.; Kausch, H.H.; Kaiser, T.; Beer, H.R. Parameters determining the strength and toughness of particulate filled epoxide resins. *J. Mater. Sci.* **1987**, *22*, 381–393. [CrossRef]
20. Maggana, C.; Pissis, P. Water sorption and diffusion studies in an epoxy resin system. *J. Polym. Sci. Part B Polym. Phys.* **1999**, *37*, 1165–1182. [CrossRef]
21. Carretti, E.; Dei, L. Physicochemical characterization of acrylic polymeric resins coating porous materials of artistic interest. *Prog. Org. Coat.* **2004**, *49*, 282–289. [CrossRef]
22. Topçuoğlu, Ö.; Altinkaya, S.A.; Balköse, D. Characterization of waterborne acrylic based paint films and measurement of their water vapor permeabilities. *Prog. Org. Coat.* **2006**, *56*, 269–278. [CrossRef]
23. Candamano, S.; Crea, F.; Coppola, L.; De Luca, P.; Coffetti, D. Influence of acrylic latex and pre-treated hemp fibers on cement based mortar properties. *Constr. Build. Mater.* **2021**, *273*, 121720. [CrossRef]
24. Shi, L.; Liu, J.; Liu, J. Influence and mechanism of polymer coating on concrete carbonization. *Dongnan Daxue Xuebao (Ziran Kexue Ban)/J. Southeast. Univ. Natural Sci. Ed.* **2010**, *40*, 208–213.
25. Toutanji, H.A.; Choi, H.; Wong, D.; Gilbert, J.A.; Alldredge, D.J. Applying a polyurea coating to high-performance organic cementitious materials. *Constr. Build. Mater.* **2013**, *38*, 1170–1179. [CrossRef]
26. Zur, E. Polyurea—The new generation of lining and coating. *Adv. Mater. Res.* **2010**, *95*, 85–86. [CrossRef]
27. McGettigan, E. Factors Affecting the Selection of Water-Repellent Treatments. *APT Bull. J. Preserv. Technol.* **1995**, *26*, 22–26. [CrossRef]
28. Ibrahim, M.; Al-Gahtani, A.S.; Maslehuddin, M.; Dakhil, F.H. Use of surface treatment materials to improve concrete durability. *J. Mater. Civ. Eng.* **1999**, *11*, 36–40. [CrossRef]
29. Moon, H.Y.; Shin, D.G.; Choi, D.S. Evaluation of the durability of mortar and concrete applied with inorganic coating material and surface treatment system. *Constr. Build. Mater.* **2007**, *21*, 362–369. [CrossRef]
30. De Muynck, W.; Debrouwer, D.; De Belie, N.; Verstraete, W. Bacterial carbonate precipitation improves the durability of cementitious materials. *Cem. Concr. Res.* **2008**, *38*, 1005–1014. [CrossRef]
31. De Muynck, W.; Cox, K.; Belie, N.D.; Verstraete, W. Bacterial carbonate precipitation as an alternative surface treatment for concrete. *Constr. Build. Mater.* **2008**, *22*, 875–885. [CrossRef]
32. Amidi, S.; Wang, J. Surface treatment of concrete bricks using calcium carbonate precipitation. *Constr. Build. Mater.* **2015**, *80*, 273–278. [CrossRef]
33. Garbacz, A.; Górka, M.; Courard, L. Effect of concrete surface treatment on adhesion in repair systems. *Mag. Concr. Res.* **2005**, *57*, 49–60. [CrossRef]
34. Soebbing, J.B.; Skabo, R.R.; Michel, H.E.; Guthikonda, G.; Sharaf, A.H. Rehabilitating water and wastewater treatment plants. *J. Prot. Coat. Linings* **1996**, *13*, 54–64.
35. Garbacz, A.; Courard, L.; Kostana, K. Characterization of concrete surface roughness and its relation to adhesion in repair systems. *Mater. Charact.* **2006**, *56*, 281–289. [CrossRef]
36. Czarnecki, L.; Garbacz, A.; Krystosiak, M. On the ultrasonic assessment of adhesion between polymer coating and concrete substrate. *Cem. Concr. Compos.* **2006**, *28*, 360–369. [CrossRef]
37. Liu, J.; Vipulanandan, C. Tensile bonding strength of epoxy coatings to concrete substrate. *Cem. Concr. Res.* **2005**, *35*, 1412–1419. [CrossRef]
38. Baltazar, L.; Santana, J.; Lopes, B.; Paula Rodrigues, M.; Correia, J.R. Surface skin protection of concrete with silicate-based impregnations: Influence of the substrate roughness and moisture. *Constr. Build. Mater.* **2014**, *70*, 191–200. [CrossRef]
39. Jia, L.; Shi, C.; Pan, X.; Zhang, J.; Wu, L. Effects of inorganic surface treatment on water permeability of cement-based materials. *Cem. Concr. Compos.* **2016**, *67*, 85–92. [CrossRef]

40. Dang, Y.; Xie, N.; Kessel, A.; McVey, E.; Pace, A.; Shi, X. Accelerated laboratory evaluation of surface treatments for protecting concrete bridge decks from salt scaling. *Constr. Build. Mater.* **2014**, *55*, 128–135. [CrossRef]

41. Franzoni, E.; Pigino, B.; Pistolesi, C. Ethyl silicate for surface protection of concrete: Performance in comparison with other inorganic surface treatments. *Cem. Concr. Compos.* **2013**, *44*, 69–76. [CrossRef]

42. Yoon, I.-S. Chloride Penetration through Cracks in High-Performance Concrete and Surface Treatment System for Crack Healing. *Adv. Mater. Sci. Eng.* **2012**, *2012*, 294571. [CrossRef]

43. Delucchi, M.; Barbucci, A.; Cerisola, G. Crack-bridging ability of organic coatings for concrete: Influence of the method of concrete cracking, thickness and nature of the coating. *Prog. Org. Coat.* **2004**, *49*, 336–341. [CrossRef]

44. Delucchi, M.; Barbucci, A.; Cerisola, G. Crack-bridging ability and liquid water permeability of protective coatings for concrete. *Prog. Org. Coat.* **1998**, *33*, 76–82. [CrossRef]

45. Delucchi, M.; Barbucci, A.; Cerisola, G. Study of the physico-chemical properties oforganic coatings for concrete degradation control. *Constr. Build. Mater.* **1997**, *11*, 365–371. [CrossRef]

46. Polder, R.B.; Borsje, H.; De Vries, H. Hydrophobic treatment of concrete against chloride penetration. *Corros. Reinf. Concr. Constr. Proc. Fourth Int. Symp.* **1996**, 183.

47. Levi, M.; Ferro, C.; Regazzoli, D.; Dotelli, G.; Presti, A.L. Comparative evaluation method of polymer surface treatments applied on high performance concrete. *J. Mater. Sci.* **2002**, *37*, 4881–4888. [CrossRef]

48. Meyer, K. Restricted Maximum Likelihood to estimate variance components for mixed models with two random factors. *Génét. Sél. Évol.* **1987**, *19*, 49. [CrossRef]

49. Dhir, R.K.; Hewlett, P.C.; Chan, Y.N. Near-surface characteristics of concrete: Assessment and development of in situ test methods. *Mag. Concr. Res.* **1987**, *39*, 183–195. [CrossRef]

50. Basheer, P.A.M.; Basheer, L.; Cleland, D.J.; Long, A.E. Surface treatments for concrete: Assessmentmethods and reported performance. *Constr. Build. Mater.* **1997**, *11*, 413–429. [CrossRef]

51. Pigino, B.; Leemann, A.; Franzoni, E.; Lura, P. Ethyl silicate for surface treatment of concrete—Part II: Characteristics and performance. *Cem. Concr. Compos.* **2012**, *34*, 313–321. [CrossRef]

52. Almusallam, A.A.; Khan, F.M.; Dulaijan, S.U.; Al-Amoudi, O.S.B. Effectiveness of surface coatings in improving concrete durability. *Cem. Concr. Compos.* **2003**, *25*, 473–481. [CrossRef]

53. Medeiros, M.; Helene, P. Efficacy of surface hydrophobic agents in reducing water and chloride ion penetration in concrete. *Mater. Struct. Constr.* **2008**, *41*, 59–71. [CrossRef]

54. Medeiros, M.H.F.; Helene, P. Surface treatment of reinforced concrete in marine environment: Influence on chloride diffusion coefficient and capillary water absorption. *Constr. Build. Mater.* **2009**, *23*, 1476–1484. [CrossRef]

55. Weisheit, S.; Unterberger, S.H.; Bader, T.; Lackner, R. Assessment of test methods for characterizing the hydrophobic nature of surface-treated High Performance Concrete. *Constr. Build. Mater.* **2016**, *110*, 145–153. [CrossRef]

56. Franzoni, E.; Varum, H.; Natali, M.E.; Bignozzi, M.C.; Melo, J.; Rocha, L.; Pereira, E. Improvement of historic reinforced concrete/mortars by impregnation and electrochemical methods. *Cem. Concr. Compos.* **2014**, *49*, 50–58. [CrossRef]

57. Buenfeld, N.R.; Zhang, J.-Z. Chloride Diffusion through Surface-Treated Mortar Specimens. *Cem. Concr. Res.* **1998**, *28*, 665–674. [CrossRef]

58. Brenna, A.; Bolzoni, F.; Beretta, S.; Ormellese, M. Long-term chloride-induced corrosion monitoring of reinforced concrete coated with commercial polymer-modified mortar and polymeric coatings. *Constr. Build. Mater.* **2013**, *48*, 734–744. [CrossRef]

59. Johannesson, B.; Utgenannt, P. Microstructural changes caused by carbonation of cement mortar. *Cem. Concr. Res.* **2001**, *31*, 925–931. [CrossRef]

60. Papadakis, V.G. Effect of supplementary cementing materials on concrete resistance against carbonation and chloride ingress. *Cem. Concr. Res.* **2000**, *30*, 291–299. [CrossRef]

61. Chang, C.-F.; Chen, J.-W. The experimental investigation of concrete carbonation depth. *Cem. Concr. Res.* **2006**, *36*, 1760–1767. [CrossRef]

62. Jiang, L.; Xue, X.; Zhang, W.; Yang, J.; Zhang, H.; Li, Y.; Zhang, R.; Zhang, Z.; Xu, L.; Qu, J.; et al. The investigation of factors affecting the water impermeability of inorganic sodium silicate-based concrete sealers. *Constr. Build. Mater.* **2015**, *93*, 729–736. [CrossRef]

63. Glasser, F.P.; Marchand, J.; Samson, E. Durability of concrete—Degradation phenomena involving detrimental chemical reactions. *Cem. Concr. Res.* **2008**, *38*, 226–246. [CrossRef]

64. Chi, J.M.; Huang, R.; Yang, C.C. Effects of carbonation on mechanical properties and durability of concrete using accelerated testing method. *J. Mar. Sci. Technol.* **2002**, *10*, 14–20.

65. Aguiar, J.B.; Júnior, C. Carbonation of surface protected concrete. *Constr. Build. Mater.* **2013**, *49*, 478–483. [CrossRef]

66. Basheer, L.; Cleland, D.J. Freeze–thaw resistance of concretes treated with pore liners. *Constr. Build. Mater.* **2006**, *20*, 990–998. [CrossRef]

67. Liu, Z.; Hansen, W. Effect of hydrophobic surface treatment on freeze-thaw durability of concrete. *Cem. Concr. Compos.* **2016**, *69*, 49–60. [CrossRef]

68. Pigeon, M.; Prevost, J.; Simard, J.-M. Freeze-thaw durability versus freezing rate. *J. Am. Concr. Inst.* **1985**, *82*, 684–692.

69. *EN 1766-Products and Systems for the Protection and Repair of Concrete Structures-Test. Methods-Reference Concrete for Testing*; ISO: Geneva, Switzerland, 2017.

70. *EN 934-2-Admixtures for Concrete, Mortar and Grout-Part 2: Concrete Admixtures-Definitions, Requirements, Conformity, Marking and Labelling*; ISO: Geneva, Switzerland, 2009.
71. *EN ISO 9514-Paints and Varnishes-Determination of the Pot Life of Multicomponent Coating Systems-Preparation and Conditioning of Samples and Guidelines for Testing*; ISO: Geneva, Switzerland, 2020.
72. *EN 1542-Products and Systems for the Protection and Repair of Concrete Structures-Test. Methods-Measurement of Bond Strength by Pull-Off*; ISO: Geneva, Switzerland, 1999.
73. *EN 13687-1-Products and Systems for the Protection and Repair of Concrete Structures-Test. Methods-Determination of Thermal Compatibility-Part. 1: Freeze-Thaw Cycling with De-Icing Salt Immersion*; ISO: Geneva, Switzerland, 2002.
74. *EN 1062-6-Paints and Varnishes-Coating Materials and Coating Systems for Exterior Masonry and Concrete-Part 6: Determination of Carbon Dioxide Permeability*; ISO: Geneva, Switzerland, 2002.
75. *EN ISO 7783-Paints and Varnishes-Determination of Water-Vapour Transmission Properties-Cup Method*; ISO: Geneva, Switzerland, 2018.
76. *EN 1062-3-Paints and Varnishes-Coating Materials and Coating Systems for Exterior Masonry and Concrete-Part 3: Determination of Liquid Water Permeability*; ISO: Geneva, Switzerland, 2008.
77. *EN 1062-7-Paints and Varnishes-Coating Materials and Coating Systems for Exterior Masonry and Concrete-Part 6: Determination of Crack Bridging Properties*; ISO: Geneva, Switzerland, 2004.
78. *EN ISO 5470-1-Rubber or Plastics-Coated Fabrics-Determination of Abrasion Resistance-Part 1: Taber abrader*; ISO: Geneva, Switzerland, 2016.
79. *EN 13529-Products and Systems for the Protection and Repair of Concrete Structures-Test. Methods-Resistance to Severe Chemical Attack*; ISO: Geneva, Switzerland, 2003.
80. *UNI 8298-8-Edilizia-Rivestimenti Resinosi per Pavimentazioni-Parte 8: Determinazione della Resistenza Alla Pressione Idrostatica Inversa*; ISO: Geneva, Switzerland, 1986. (In Italian)
81. *EN 13501-1-Fire Classification of Construction Products and Building Elements-Part 1: Classification Using Data from Reaction to Fire Tests*; ISO: Geneva, Switzerland, 2018.
82. *EN 1062-11-Paints and Varnishes-Coating Materials and Coating Systems for Exterior Masonry and Concrete-Part 11: Methods of Conditioning before Testing*; ISO: Geneva, Switzerland, 2002.
83. Barbucci, A.; Delucchi, M.; Cerisola, G. Organic coatings for concrete protection: Liquid water and water vapour permeabilities. *Prog. Org. Coat.* **1997**, *30*, 293–297. [CrossRef]
84. Safiuddin, M.; Soudki, K.A. Water vapor transmission and waterproofing performance of concrete sealer and coating systems. *J. Civ. Eng. Manag.* **2015**, *21*, 837–844. [CrossRef]
85. Aguiar, J.B.; Camões, A.; Moreira, P.M. Performance of concrete in aggressive environment. *Int. J. Concr. Struct. Mater.* **2008**, *2*, 21–25.
86. *EN ISO 868-Plastics and Ebonite-Determination of Indentation Hardness by Means of a Durometer (Shore Hardness)*; ISO: Geneva, Switzerland, 2003.
87. Redner, J.A.; Hsi, R.P.; Esfandi, E.J. Evaluating coatings for concrete in wastewater facilities. An update. *J. Prot. Coat. Linings* **1994**, *11*, 50–61.
88. Reddy, B.; Sykes, J.M. Degradation of organic coatings in a corrosive environment: A study by scanning Kelvin probe and scanning acoustic microscope. *Prog. Org. Coat.* **2005**, *52*, 280–287. [CrossRef]
89. De Muynck, W.; De Belie, N.; Verstraete, W. Effectiveness of admixtures, surface treatments and antimicrobial compounds against biogenic sulfuric acid corrosion of concrete. *Cem. Concr. Compos.* **2009**, *31*, 163–170. [CrossRef]
90. *EN ISO 1182-Reaction to Fire Tests for Products-Non-Combustibility Test*; ISO: Geneva, Switzerland, 2020.

 *materials*

Article

Evaluation of the Protection Ability of a Magnesium Hydroxide Coating against the Bio-Corrosion of Concrete Sewer Pipes, by Using Short and Long Duration Accelerated Acid Spraying Tests

Domna Merachtsaki [1], Eirini-Chrysanthi Tsardaka [2], Eleftherios Anastasiou [2] and Anastasios Zouboulis [1],*

[1] Laboratory of Chemical and Environmental Technology, Department of Chemistry, Aristotle University of Thessaloniki, 54124 Thessaloniki, Greece; meradomn@chem.auth.gr

[2] Laboratory of Building Materials, Department of Civil Engineering, Aristotle University of Thessaloniki, 54124 Thessaloniki, Greece; extsardaka@gmail.com (E.-C.T.); elan@civil.auth.gr (E.A.)

* Correspondence: zoubouli@chem.auth.gr

Abstract: The Microbiologically Induced Corrosion (MIC) of concrete sewer pipes is a commonly known problem that can lead to the destruction of the system, creating multiple public health issues and the need for costly repair investments. The present study focuses on the development of a magnesium hydroxide coating, with optimized properties to protect concrete against MIC. The anti-corrosion properties of the respective coating were evaluated by using short and long duration accelerated sulfuric acid spraying tests. The coating presented satisfying adhesion ability, based on pull-off and Scanning Electron Microscopy (SEM) analysis measurements. The surface pH of the coated concrete was maintained at the alkaline region (i.e., >8.0) throughout the duration of all acid spraying tests. The consumption of the coating, due to the reaction (neutralization) with sulfuric acid, was confirmed by the respective mass and thickness measurements. The protection ability of this coating was also evaluated by recording the formation of gypsum (i.e., the main corrosion product of concrete) during the performed tests, by X-ray Diffraction (XRD) analysis and by the Attenuated Total Reflectance (ATR) measurements. Finally, a long duration acid spraying test was additionally used to evaluate the behavior of the coating, simulating better the conditions existing in a real sewer pipe, and the obtained results showed that this coating is capable of offering prolonged protection to the concrete substrate.

Keywords: concrete bio-corrosion; sulfuric acid corrosion control; magnesium hydroxide coating; sewerage pipe systems; acid spraying test

Citation: Merachtsaki, D.; Tsardaka E.-C.; Anastasiou, E.; Zouboulis, A. Evaluation of the Protection Ability of a Magnesium Hydroxide Coating against the Bio-Corrosion of Concrete Sewer Pipes, by Using Short and Long Duration Accelerated Acid Spraying Tests. *Materials* **2021**, *14*, 4897. https://doi.org/10.3390/ma14174897

Academic Editor: Frank Collins

Received: 21 July 2021
Accepted: 27 August 2021
Published: 28 August 2021

Publisher's Note: MDPI stays neutral with regard to jurisdictional claims in published maps and institutional affiliations.

1. Introduction

The good operation and efficient protection of sewerage pipeline systems, especially against MIC, is majorly important for urbanized societies. MIC is a specific type of corrosion mechanism that can take place usually in the larger diameter sewer pipes, constructed mainly from concrete, expected to highly affect the concrete and reduce the lifetime of the structure. This phenomenon has been widely studied during the last 100 years [1–5] and the scientific community has been focusing on the application of different mitigation techniques in order to block the collapse of the sewerage system and to prevent the costly repair/replacement of affected sewer pipes [6].

MIC is a complex corrosion mechanism, depending on the multiple chemical and biochemical processes which can take place in the sewer pipeline. Sulfates, along with other substances carried within wastewater, initiate the growth of microbial communities, which in turn cause the reduction of sulfates to hydrogen sulfide. The formed hydrogen sulfide (gas) is released to the liquid phase (wastewater) and finally, it is emitted in the upper (air) part of the (usually semi-filled) pipe. Then, the gas hydrogen sulfide can

143

be dissolved in the humidity film located on the upper region of the pipe wall (usually denoted as "crown") [7,8]. As a result, the pH of the concrete's surface becomes less alkaline, and at the critical pH value of 9, Neutrophilic Sulfur Oxidizing Bacteria (NSOB) can be developed, using the hydrogen sulfide to produce biogenic sulfuric acid [9,10]. As a result the surface pH of concrete further lowers towards acidic values, until the development of Acidophilic Sulfur Oxidizing Bacteria (ASOB) at pH value 4 [10,11]. The development of these microorganisms onto the inner concrete pipe surface can lead to the production of additional biogenic sulfuric acid. The extended formation of biogenic sulfuric acid is crucial for the overall structure's preservation and stability, because the alkaline components of concrete are reacting with acid and being consumed, forming corrosion by-products, such as gypsum and ettringite [8,10]. Finally, the concrete degrades further and the construction collapses, leading (among others) to major odor and pollution problems (e.g., leaking wastewater, polluting soil, etc.) [12,13].

The prevention methods developed so far may include the application of linings and coatings [14–16], located onto the inner side of sewer pipes, as well as the use of new/improved pipe materials with intrinsic anti-corrosion properties [17,18], or the appropriate chemical dosing of specific agents directly in the wastewater stream [19,20]. According to the relevant literature linings and coatings present some advantages in respect to other mitigation technologies. More specifically, coatings can create a barrier between the concrete sewer and the corrosive environment, hence offering a good protection. Additionally, the application of coatings is relatively cheaper, when compared to the replacement of destructed sewer pipes with new pipes. Finally, the environmental concerns that the alternative protection method of chemical dosing with different reagents may lead to, seem to be avoided when lining and coatings are used [21,22].

Certain coatings and linings have been already widely used for the protection of concrete [15,16,23] or steel structures (such as manganese hydrate coatings) [24]. These methods can protect the concrete surface either by blocking the interaction of substrate with the corrosive environment, or by reacting (e.g., neutralization) with the corrosive substances. Multiple protective coatings have been used for protection against MIC, according to the relevant literature, such as polymer-based (epoxy, polyurethanes), inorganic coatings (magnesium hydroxide coatings, alkali-activated materials), or cementitious coatings modified with polymers [25]. All these coatings may present specific advantages and disadvantages regarding their performance, by offering different kinds/types of protection [10]. Polymeric-based coatings isolate the concrete from the aggressive environment, without, however, reacting to or neutralizing the biogenic produced acid. Magnesium hydroxide coatings can neutralize the biogenic sulfuric acid, and can diminish it in the sewerage environment; therefore, the acid cannot initiate any secondary kind of corrosion, as compared to the polymeric-based coatings, where the acid continues to exist in the system [26,27]. The main properties of magnesium hydroxide, when applied as corrosion protection coating, are its ability to maintain for sufficient time the alkaline surface pH values of concrete substrate, hence blocking the development of sulfur oxidizing bacteria, and its ability to react (neutralize) with the produced biogenic sulfuric acid in case of microorganisms' development.

This study is mainly focused on a specific magnesium hydroxide (MH) slurry, which after preliminary experiments was found to present the optimal expected properties, e.g., regarding the specific surface area (SSA) and the particle size distribution (PSD) [28]. The magnesium hydroxide is evaluated mainly regarding its action as protective coating on concrete substrates, under conditions that simulate those existing in the sewer pipes. The coating's properties were additionally evaluated by examining the formation of gypsum as the main concrete corrosion by-product. The aim of this paper was to confirm that the initial physicochemical properties of used MH can affect and enhance the properties of the protective coating. The originality of this research is based on the use of a long duration acid spraying test (for four months), which is closer to the real conditions existing in a sewer pipe, especially when compared with the accelerated tests of smaller duration.

Moreover, the examined concrete specimens were also replaced by proper poly(methyl methacrylate) (PMMA) substrates in order to isolate the possible side effects of concrete presence and to study the coating's behavior under the acidic environment conditions (i.e., reacting alone with sulfuric acid).

2. Materials and Methods

2.1. Substrates

As far as concrete specimens are concerned, concrete type MC (0.45) was produced, according to the EN 1766:2017 Standard [29], conforming to the requirements of the EN 1916:2002 Standard [30], aiming the substrate to resemble the corresponding concrete materials, used for the construction of sewer pipes. The water/cement ratio was 0.45 and the proportions of the used constituents were: 410 kg/m^3 CEM I 42.5 R, 895 kg/m^3 crushed limestone sand (0–4 mm), 895 kg/m^3 crushed limestone aggregates (4–8 mm), 184.5 kg/m^3 tap water and 0.5% wt. of cement poly-carboxylate based super-plasticizer. Wooden molds with different dimensions (i.e., 200 mm × 200 mm × 20 mm and 50 mm × 50 mm × 20 mm), depending on each test specific requirements, were used for the preparation of specimens. After 24 h the concrete specimens were de-molded and cured into water for 27 days at 20 ± 2 °C.

The use of concrete specimens, as substrates for the examined protective coating, may obstruct the observation of the coating's consumption, occurred due to the potential (neutralization) reaction of sulfuric acid with the cement (alkaline) paste of concrete. Therefore, in this study, poly (methyl methacrylate) (PMMA) plates were alternatively used as substrates, apart from the concrete substrates, in order to promote a clear observation of the coating reaction with the sulfuric acid. PMMA plates with dimensions of 50 mm × 50 mm × 5 mm were cut and used for the detailed examination of coating behavior under similar (environmental) conditions.

2.2. Surface Coatings

2.2.1. Magnesium Hydroxide Slurry

The magnesium hydroxide slurry was provided by Grecian Magnesite S.A. It was produced in the laboratory by the controlled hydration of micro-crystalline Caustic-Calcined Magnesia (CCM), of medium purity. The MH slurry has a solids content of 57.5%, specific gravity of 1.46 g/cm^3, a viscosity of 2500 cP (R5 spindle @100 rpm), and contains 0.4% on solids of a modified methyl cellulose as convenient adhesive material.

The main physicochemical characteristics of magnesium hydroxide solids of the slurry are provided in Table 1. The total mass loss at 1000 °C (i.e., the Loss on Ignition, LOI) corresponds to all the water and CO_2 content of the respective powder.

Table 1. The main physicochemical characteristics of used slurry solids.

Nominal Chemical Composition					Mg(OH)$_2$ Content (%)	SSA (m^2/g)	PSD (µm)	
MgO (%)	SiO$_2$ (%)	CaO (%)	Fe$_2$O$_3$ (%)	LOI (%)			d$_{50}$	d$_{90}$
62.81	4.25	2.46	0.25	30.11	89.0	7	3.8	13.1

This magnesium hydroxide differs mainly in the MgO content, but also in the specific surface area and in the particle size from other examined magnesium hydroxides, previously investigated [28]. The main difference regards the specific surface area, which is very low in this case (7 m^2/g) to maintain favorable slurry characteristics, such as viscosity and workability and to allow the inclusion of cellulose. The results of the previous study [28] indicate that the bigger particle size may lead to slower interactions with the respective (acidic) environment. In that way, this magnesium hydroxide has finer particles, than other examined relevant hydroxides [28]. Additionally, the purity of raw CCM used to prepare the studied slurry, was higher, and according to previous results [28], this fact may enhance the preservation of alkaline pH values in the protected surface after the application of this coating.

2.2.2. Coating Application

The thickness of relevant protective coatings, according to the literature and preliminary testing, can range from few micrometers to few millimeters, depending on the coating's nature [23,31,32]. In this study the thickness was selected to be between 1.0–1.5 mm and was designated according to the specific amount of applied coating, i.e., 0.0018–0.0020 g/mm^2. After the application of coating onto the substrates, the specimens were dried for 3 days under normal laboratory conditions (i.e., 21 ± 2 °C and relative humidity 60 ± 10%) before further testing.

The examined coating was also evaluated for its adhesion ability onto the concrete, by applying the pull-off bond testing method, according to the standards EN 1542:1999 [33] and EN 13578:2003 [34]. The coating was applied on concrete specimens with dimensions of 200 mm × 200 mm × 20 mm in order to perform the respective measurements. According to these standards, the pull-off equipment (digital pull-off strength tester, Matest) was used in order to record the failure load, as well as the specific type of failure.

2.3. Scanning Electron Microscopy Analysis

The adhesion of coating onto the concrete surfaces was also examined by Scanning Electron Microscopy (SEM) analysis, providing a closer examination of the respective concrete-coating interface and of the coating's morphology. Moreover, specific information regarding the magnesium content, which might penetrate the concrete structure, can be obtained by using the EDS analysis. The specimens were cut vertically in order to reveal the interface between the concrete and the coating and were polished appropriately, creating a flat surface for clear observation. Micrographs were obtained by using a JSM-7610F Plus, JEOL, scanning electron microscope and the EDS analysis was performed by using the AZtec Energy Advanced X-act System (Oxford Instruments, Oxfordside, UK).

2.4. Accelerated Spraying Tests

According to a relevant study [28] two different accelerated sulfuric acid spraying tests were applied, so that coatings can be properly evaluated in terms of the protection they can offer to concrete surfaces in relatively short time period. The same tests and conditions were also used in this study in order to examine the selected magnesium hydroxide, presenting optimized properties. In particular, the coating was firstly examined on concrete substrates, by using a hand-held spraying device (HHD), which prayed sulfuric acid onto the coated concrete surface. Then, the coating was examined, applied onto concrete specimens in a custom-made spraying laboratory chamber, as described in [28]. In both tests, the coated specimens were placed vertically in order to simulate the real pipe walls. The laboratory chamber was constructed from poly (methyl methacrylate) and it was equipped with nebulizers and appropriate supports placed opposite them, in order to place the concrete specimens vertically. Additionally, the respective coating was evaluated in the acid spraying chamber, by applying it onto the PMMA plates, in order to evaluate the protecting properties of the coating, by eliminating the side effects of corroded concrete substrate. Finally, the spraying chamber was also used for a longer duration acid spraying test, so that the durability and effectiveness of this coating could be examined under more realistic conditions, using daily a lower acid amount.

For the application of the coating onto the concrete substrate, concrete specimens with dimensions of 50 mm × 50 mm × 20 mm were used for the acid spraying tests. Moreover, PMMA plates with the same dimensions were also comparatively used for the respective tests.

The surface pH values, the mass change and the existing mineralogical phases were recorded for all performed tests. However, the thickness change of the coating was examined only after the HHD test. The spraying tests and the measurements applied for each case are summarized in Table 2.

Table 2. The acid spraying tests and the respective measurements.

Test	Measurements
4 days hand-held device test	Surface pH Thickness Mass Mineralogical phases
4 days spraying chamber test (concrete specimens)	Surface pH Mass Mineralogical phases
4 days spraying chamber test (PMMA plates)	Surface pH Mass Mineralogical phases
128 days spraying chamber test	Surface pH Mass Mineralogical phases

In all acid spraying tests a flat surface pH electrode (Extech PH100: Waterproof ExStik pH meter, Extech Instruments, Nashua, NH, USA) was used to record the surface pH values of specimens. The initial surface pH values were recorded for each specimen before the beginning of acid spraying tests. The surfaces to be measured were wetted with 1 mL of deionized water prior to the measurement.

An electronic balance Kern PCB 350–3 (350 $\pm$ 0.001 g) was used to record the mass change of specimens daily, in order to evaluate the consumption of the coating. Firstly, the initial weight of all uncoated concrete specimens was recorded, as well as their weight after the application of the slurry, in order to calculate the mass of dry coatings.

A Dino Lite Digital Microscope was used to photograph the cross-section of coated concrete specimens before and after the application of acid spraying, in order to record the thickness change of the coating. After that, the photographs were digitized and processed by using the microscope's software and finally, the thickness was calculated according to the magnification used in each picture.

The mineralogical phases were evaluated by using the X-ray diffraction analysis (XRD), which is described in Section 2.5.

2.4.1. Stoichiometry Calculations

Sulfuric acid solutions were used in all performed tests, aiming to simulate the biogenic sulfuric acid produced in a sewer pipe internal wall. The sulfuric acid spraying tests were performed to examine the total consumption of magnesium hydroxide coating at the end of respective tests, due to the neutralization reaction between them:

$$Mg(OH)_2 + H_2SO_4 \rightarrow MgSO_4 + 2H_2O \tag{1}$$

In that way, the amount of daily sprayed sulfuric acid (according to the reaction's stoichiometry) was calculated, based on the amount of applied coating, on the duration of each test and on the used acidic solution concentration.

The duration of accelerated acid spraying tests was selected to be 4 days, apart from the long duration test, which lasted 128 days. The used sulfuric acid solutions had concentrations of 4 M, 0.2 M and 0.1 M for the HHD spraying test, the 4-day chamber test and the 4-month chamber test, respectively. In all these experiments the same conditions were applied for the calculation of corresponding necessary sulfuric acid amount to be sprayed, as explained in more detail in the relevant literature [32].

The calculated stoichiometry (i.e., the moles of sulfuric acid) was divided by the number of spraying days and the daily spraying application stoichiometry corresponded to 25% of the overall stoichiometry, regarding the respective neutralization reaction between the coating and the sulfuric acid for the 4-day tests (i.e., the HHD spraying test and the chamber spraying tests). The daily spraying applications were different between these two tests, due to the different initial acid solutions concentrations used. Accordingly, the daily spraying application stoichiometry for the longer duration test (128 days) corresponded to the 0.78% of the overall stoichiometry of the respective reaction.

2.4.2. 4-Days Accelerated Acid Spraying Test

Two tests were used in the case of 4-days accelerated acid spraying tests, i.e., the test where the hand-held device was used and the test where the custom-made poly (methyl methacrylate) chamber was used. The concentration of sulfuric acid solution in the HHD test was 4 M, in order to magnify the consumption of coating, due to the reaction with sulfuric acid, whereas the concentration of the solution in the chamber tests was 0.2 M, in order to approach the milder conditions commonly existing in a sewer pipe.

Coated concrete specimens were used in both tests, whereas the coated PMMA plates were only tested in the spraying chamber test. The coating was dried after its application for 3 days under normal laboratory conditions before the sulfuric acid spraying applications. Four coated specimens were removed daily from each acid spraying procedure and studied in detail.

2.4.3. 4-Months Accelerated Acid Spraying Test

The long-term durability of the examined coating was evaluated by a 4-month accelerated acid spraying test that was held in a custom-made poly (methyl methacrylate) spraying chamber. The rinsing liquids were collected separately for each specimen (chamber facility) in order to be further examined. The concentration of sprayed sulfuric acid (0.1 M) was selected in this case to be lower in respect to the 4-day tests (0.2 M), in order to simulate better the conditions existing in a sewer pipe. The longer duration test allowed the spraying of significantly lower amount of sulfuric acid daily (according to the number of daily spraying applications). In that way, the behavior and consumption of coating in time can be examined.

At 8-days intervals, four specimens were removed from the chamber and tested, regarding the surface pH values, the mass change and the mineralogical phases existing on the surface. Moreover, the rinsing liquids were selected during the same time periods, in order to determine the magnesium content that was rinsed/removed (either as product, or as leaching) during the experiment. The magnesium content was determined by Flame Atomic Absorption Spectrophotometry, using the Perkin-Elmer AAnalyst 400 instrument. The experimental conditions, applied in the chamber, simulated the usual sewer pipe conditions, i.e., maintaining the temperature at 20 ± 2 °C and 99% relative humidity.

2.5. X-ray Diffraction (XRD) Analysis and Attenuated Total Reflectance (ATR)

The recognition and quantitative determination of mineralogical phases existing on the specimens' surface were performed with the XRD analysis. After the spraying process, the specimens were dried at 40 °C for 24 h, and then the coating and the potentially formed by-products were scratched from the top of specimens, ground and measured. The structural phases (mineralogical composition) of the obtained samples were analyzed by XRD measurements, using a PW 1840 Phillips diffractometer with CuKa radiation, step size of 0.02° and step time of 0.4 s, operating at 30 kV and 10 mA. The obtained diffractograms were further quantified by following the Rietveld methodology, using the FullProf Suite Software.

ATR was additionally used for the characterization of selected surface samples from the longer duration test. For the ATR measurements a Cary 630 FTIR, Agilent Technologies and the respective software MicroLab were used. The samples were measured directly after the XRD measurements without any prior treatment.

3. Results

3.1. Adhesion Measurements

The results of pull-off measurements can offer important information about the adhesion ability of protective coatings onto an appropriate substrate. In particular, the coating's tensile bond strength can be calculated by using the failure load, resulting from the respective measurements. Additionally, the type of failure can be determined optically, according to the aforementioned standards [33,34] and expressed as a percentage, based on the relevant surface area. Two concrete specimens, with 4 testing areas on each one, were used for the evaluation of each sample coating.

According to the standards, and regarding the pull-off measurements, several types of coating failures can be optically observed and defined, depending on the applied coatings. However, in this study, two specific types of failure were present, i.e., the adhesion failure between the substrate/concrete and the coating (denoted hereafter as A/B), and the cohesion failure within the layer of coating itself (denoted hereafter as B). The results, regarding these failure types, are usually presented as a combination between the two aforementioned types (and as percentage).

The tensile bond strength between the applied coating and the concrete substrate, as well as the type of failure, is presented in Table 3. The results of studied coating are compared with the results of other relevant coatings, previously examined [28], in order to draw useful conclusions and are commented below.

Table 3. Tensile bond strength (f_h) values, the respective standard deviation (SD) and the specific type of coatings' failure.

f_h (MPa)	SD (MPa)	Type of Failure	
		A/B (%) [1]	B (%) [2]
0.30	0.025	30	70

[1] Adhesion failure between the substrate and the coating layer; [2] Cohesion failure within the layer of coating.

As far as the tensile bond is concerned, the coating was found to present medium-to-high values when compared to other relevant coatings, consisting of different magnesium hydroxide powders [28]. The specific type of failure plays a key role in understanding the adhesion ability of coating, when combined with the tensile bond strength. The application of examined coating generally requires high values of the B failure type, so that in case of coating detachment, a significant amount of coating can continue to protect the substrate. Therefore, the calculated values, regarding the two types of failure, are considered as satisfactory, because the coating exhibited more type-B failure (70%), than type-A/B (30%). As a result, this coating combined the required type of failure with a satisfactory tensile bond strength value, as compared with other relevant magnesium hydroxide coatings [28].

3.2. SEM Analysis

Figure 1 is a cross-section micrograph of the coated concrete specimen, where the coating is presented in the upper side, whereas the concrete substrate is presented in the lower side of the image. The adhesion ability of the coating can be verified by this figure, because the coating seems to be firmly attached onto the concrete surface, without any voids or cracks observed at the interface area. The concrete surface is distinct, because it presents different roughness and porosity from the coating. The structure of concrete is denser than the respective of coating, which seems to be rather porous. It is important to note that the selected concrete type is of relatively low porosity. Additionally, the scratch polishing lines are rougher on the concrete's surface, than on the coating's surface. The coating thickness ranged between 0.4 and 0.7 mm, according to these observations. A closer look at the coating-concrete interface (Figure 1, right) confirms the good adhesion of the coating onto the surface.

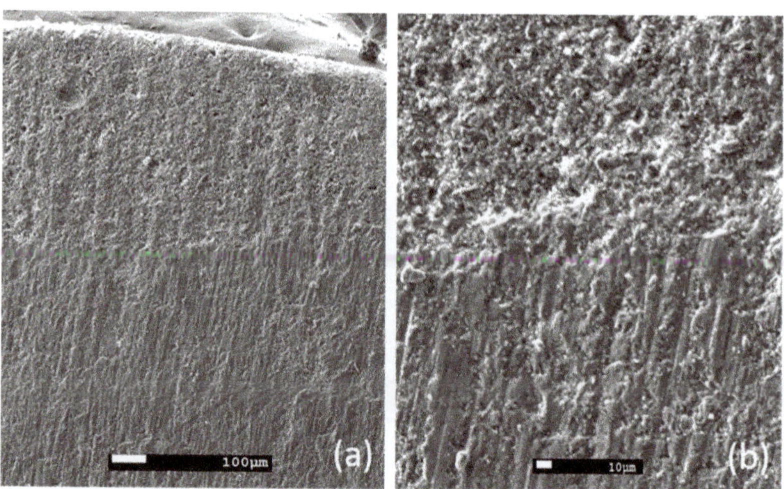

Figure 1. SEM micrograph of the coated concrete specimen at the interface region: magnification (**a**) ×60, (**b**) ×200.

As far as the coating's microstructure is concerned (Figure 2), the particles (magnesium hydroxide particles, or magnesium oxide) are mainly spherical and elongated, usually presenting a size around 1 μm. The structure is porous and loose, while the EDS analysis showed mainly the presence of magnesium (apart from oxygen).

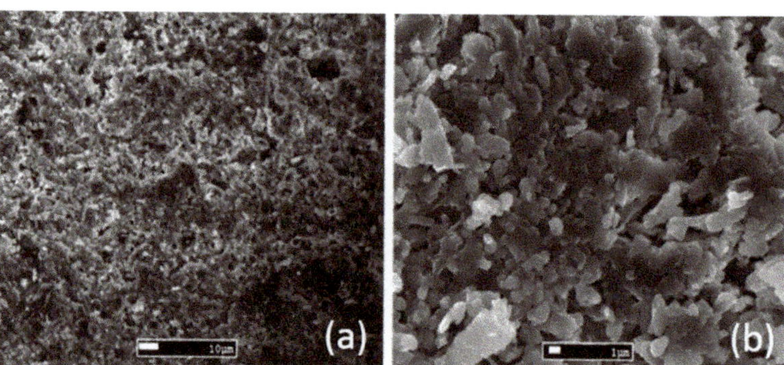

Figure 2. SEM micrographs regarding the structure of examined coating: magnification (**a**) ×500, (**b**) ×3000.

The EDS analysis of the coating and of the concrete surface can provide information regarding the potential penetration of coating constituents into the concrete structure, which can also explain the coating's good adhesion and create an additional protection zone in the concrete structure. The EDS analysis of the spectrum areas (Figure 3) is presented in Table 4. The results showed that the magnesium content in the concrete's structure (and in the depth of ~100 μm) is relatively small (1.41% wt.). According to the relevant literature [28], this corresponds to the initial magnesium content of concrete. In contrast to the previous study, this coating does not seem to penetrate the concrete substrate. The increased calcium content of concrete surface (Spectrum 2) is attributed to the calcium silicate compounds of the hydrated cement paste, as well as to the limestone (calcium carbonate) source of aggregates.

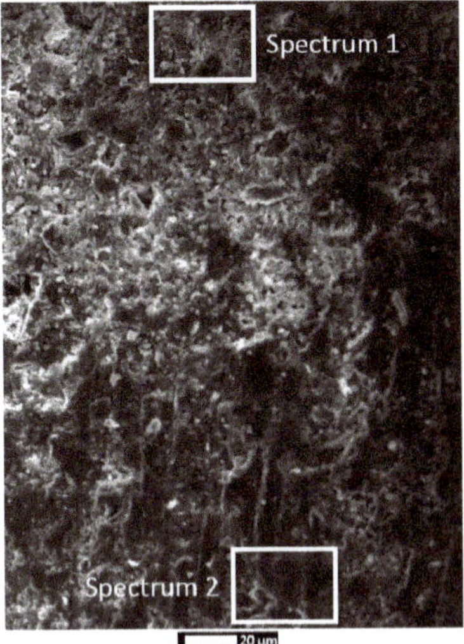

Figure 3. SEM micrograph of the coating-concrete interface, showing the respective energy-dispersive X-ray spectroscopy (EDS) spectral areas.

Table 4. EDS spectra of the coating and of the concrete, as presented in Figure 3.

Elements	Weight %	
	Spectrum 1	Spectrum 2
Mg	94.16	1.41
Si	2.31	1.24
Ca	3.54	97.35

3.3. Accelerated Sulfuric Acid Spraying Tests

3.3.1. 4-Days Accelerated Acid Spraying Tests

The study of coating surface pH values during its interaction with sulfuric acid is very important, because this is the main acquired and inherent property of magnesium hydroxide coating, against the biogenic corrosion of concrete. The maintenance of alkaline surface pH values can block the development of sulfur oxidizing bacteria, hence the production of biogenic sulfuric acid and the subsequent corrosion of concrete.

The daily recordings of surface pH values, regarding the coated concrete specimens, when sprayed during the HHD test, as well as the coated concrete specimens sprayed in the spraying chamber are presented in Figure 4. Moreover, the surface pH values of the coated PMMA plates are also presented in Figure 4.

In all these cases, the starting pH values of the coating, measured on the substrates' surface, were very similar. The coating presented initial surface pH values around value 10 (day 0). Despite the differences between the tests, the application of acid spraying led to slightly decreased surface pH values (i.e., 8.00–9.25), comparing with the initial pH values. However, the pH values remained in the alkaline region throughout the spraying processes, which means that the coating can block effectively the development of bacteria. The differences of pH values between the tests found to be relatively small. More precisely, after four days of spraying, the values measured for the specimens, sprayed under the HHD test, were the highest, i.e., 9.25. This fact may be due to the smaller number of spraying applications in the case of HHD test, even though a highly concentrated sulfuric acid solution was used, as compared with the number of spraying applications in the case of the chamber test.

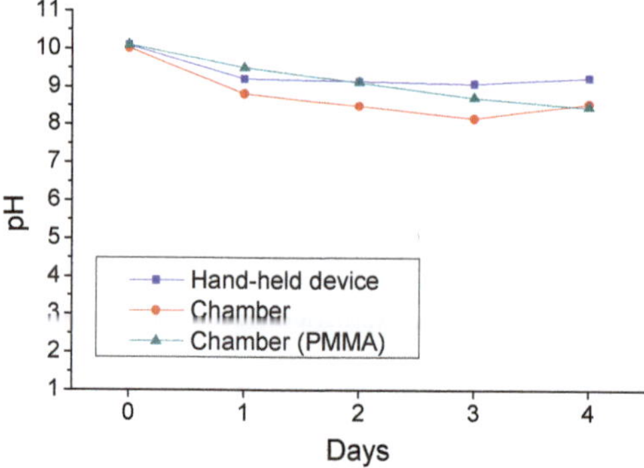

Figure 4. Surface pH values of coated concrete and PMMA specimens, during the application of different acid spraying tests.

The respective pH values of coated PMMA plates were found to be slightly higher, at days 1, 2, and 3 of chamber spraying test, with respect to the values of coated concrete specimens. The comparison of surface pH values for the coating, during the chamber test, as applied either on concrete substrates (dry coating), or on PMMA plates, indicates that the presence of concrete substrate might also influence the measured pH values.

These values can be compared with the corresponding ones of other magnesium hydroxide coatings, as well as of uncoated concrete specimens, which were sprayed accordingly [28]. The examined coating exhibited satisfying surface pH values (pH > 8) in most cases of spraying tests, although the initial pH value of it at day 0 was not the highest among the several relevant coatings.

Moving on, mass measurements were also conducted for the coated concrete specimens subjected to both tests (HHD and chamber), as well as for the coated PMMA plates (chamber test). The results are presented as percentage, in respect to the initial coating mass and are given in Figure 5.

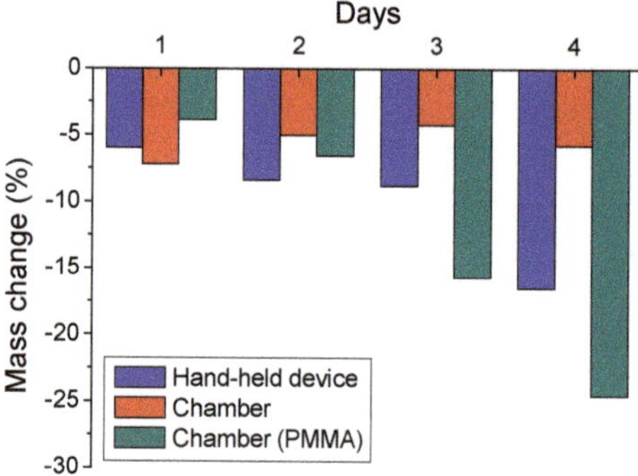

Figure 5. Mass changes of the coated concrete specimens after the application of HHD spraying test, as well as of the coated concrete and PMMA plates after the application of chamber spraying test.

According to the stoichiometry calculations, a larger consumption of the coating, at all these cases, can be expected, because the amount of sulfuric acid that was sprayed onto the specimens, during the 4-days tests, corresponded to the coating's total magnesium hydroxide amount. However, it can be assumed that the sprayed sulfuric acid can only react with the exposed coating surface, instead of the full coating's mass [28], somehow reducing the mass consumption respectively.

The mass change results indicated that the substrate affects the behavior of the coating concerning mass change. The mass change of the coating on the concrete specimens (chamber test) presented rather stable behavior during the spraying procedure. On the other hand, the mass change of coating on the PMMA specimens was more intensified during the test. In this case the mass decrease reached almost 25%, when compared to the initial mass of the system. It seems possible that the adhesion between the coating and the concrete can play an important role on the system's behavior. In this respect the porosity of concrete and of concrete surface are important parameters. On the contrary, the absence of adhesion between the used coating and the polymer substrate (PMMA) did not contribute to that direction.

The concentration of sprayed sulfuric acid can also play an important role, as during the HHD test (using the higher concentration of 4 M), the mass loss was greater than in the chamber test (using the lower sulfuric acid concentration of 0.1 M). This reasonable result is connected to the greater consumption of coating in the case of greater acid concentration sprayed.

It is also a fact that mass increase was not observed, as it was recorded for some other magnesium hydroxide coatings, presenting similar characteristics [28]. This is connected to the washing out rate of reaction (neutralization) by-products and the ease that the particular coating behaves under the applied spraying conditions, as its behavior depends mainly on its specific physicochemical characteristics, such as the smaller particle size and the higher surface area in comparison with other relevant materials, previously examined [28].

The measurement of coating thickness during the HHD acid spraying test was performed as another way to evaluate the coating consumption under the application of extreme acidic conditions (high concentration of sprayed sulfuric acid). The coating thickness was expected to decrease during the spraying procedure, due to its reaction with sulfuric acid. The results presented in Figure 6 confirmed that the coating thickness is gradually reduced, due to the coating's consumption. In this case the thickness reduction of studied coating was higher (−66%), than the thickness reduction of other studied relevant coatings [28]. This fact indicated that the coating was more effectively consumed, also confirming the mass change results. Additionally, the mass loss, intensified at day 4 of the experiment, agreed with the extended decrease of thickness at the same day.

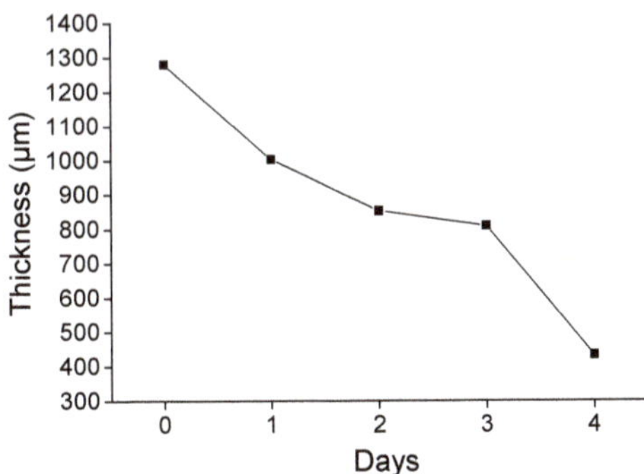

Figure 6. The evolution of thickness coating throughout the HHD spraying test.

3.3.2. 4-Months Accelerated Acid Spraying Test

The 4-months accelerated acid spraying test was performed to better simulate the existing conditions in a sewer, as described in Section 2.4.3. The surface pH, the mass change, and the mineralogical phases of surface were recorded. Figure 7 presents the results of surface pH recordings of the sprayed specimens at 8-day intervals. The surface pH values were maintained at the alkaline region throughout the long duration test and confirmed the ability of the coating to block the development of acidophilic sulfur oxidizing bacteria. In that way, only the neutrophilic sulfur oxidizing bacteria might be developed and produce a small amount of biogenic sulfuric acid. Nevertheless, based on the previous results (Section 3.3.1), the coating can still react with the acid and protect the concrete surface.

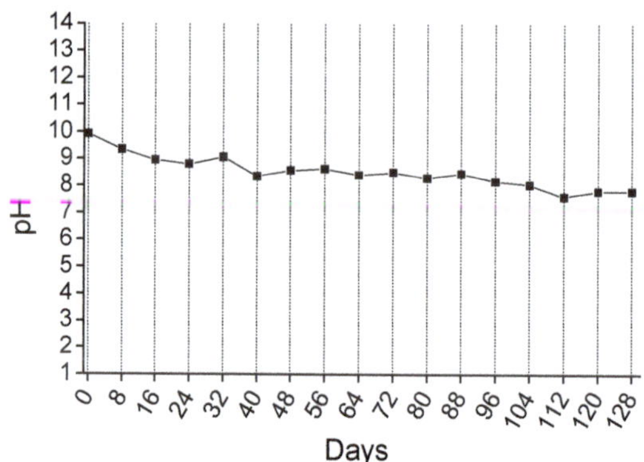

Figure 7. The evolution of surface pH values of coated concrete specimens, during the application of the long-term duration acid spraying test.

The variation of pH values during the 128 days experiment were quite similar to those obtained from the 4-day testing, considering the respective acid concentrations. This shows that the short time tests are also reliable and both tests (long- or short-term) can be performed to evaluate the pH behavior of this magnesium-based coating.

Moving on, the mass change results of the long duration acid spraying test are presented in Figure 8. The results are shown in respect to the initial coating mass (%) applied on the concrete specimens. The complicated mechanisms that took place during the spraying applications, led to rather controversial mass change results. Two different processes can take place during the acid spraying test, i.e., the reaction (neutralization) of the coating with the sprayed acid and the formation of by-products (mainly hexahydrite and gypsum). The first process leads to mass decrease, while the second one to mass increase. It is also important to note that both processes can take place simultaneously.

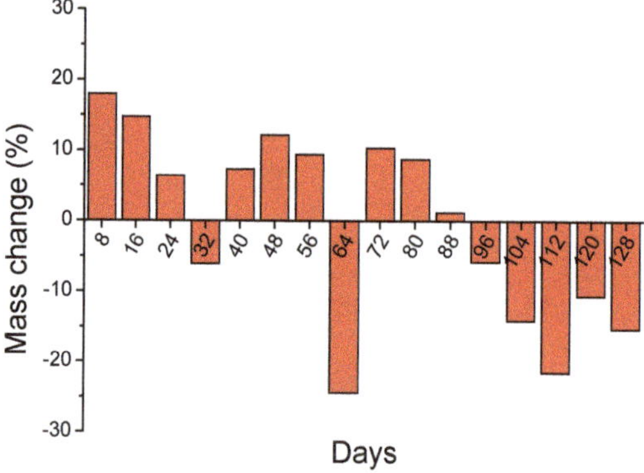

Figure 8. The evolution of mass changes of coated concrete specimens during the application of long-term duration acid spraying test.

According to the results, the mass of specimens increased (positive values) during the first 80 days, with the exceptions of specimens at day 32 and day 64. Regarding the results of day 32 specimens, there was an unregular mass decrease, with respect to the trend of the previous and the following specimen results. However, the results are given for different specimens each time, and this can lead to different results in some cases. Regarding the results of day 64, there was a technical problem during the test and mass loss was observed, due to coating detachment. After 88 days of acid spraying, mass decrease was recorded until the end of the procedure.

These results, and especially the mass increase, differed from the respective tests of the accelerated acid spraying tests (Figure 5), where no mass increase was observed. However, other magnesium hydroxide coatings presented also mass increase, during the accelerated acid spraying test, which was attributed to the formation of by-products [28]. It is a fact that the experimental conditions of the 4-months test were different, i.e., the solution concentration and the duration and the number of daily spraying applications. Accordingly, these factors enabled the long-term milder condition testing, which allowed the recording of initial increase of mass, due to the formation of by-products and the following decrease of mass, due to the consumption/degradation of coating. Based on that, both mechanisms took place. The possible formation of by-products was further examined using the XRD measurements (Section 3.4.2).

The long-term acid spraying test was also conducted in a poly (methyl methacrylate) chamber, which was modified in a way that the washouts could be collected and analyzed. After that, the magnesium content was determined in them, and the results are presented in Figure 9. The presence of magnesium in the rinsing liquids indicated the continuous reaction of magnesium hydroxide with the sprayed acid or its possible leaching. The magnesium content results seem to agree with the mass change results, because, in general, when mass decrease was observed (e.g., at day 112), there was an increase of magnesium content in the rinsing liquids. However, magnesium was detected, even when the mass increase was recorded. This indicates that an amount of magnesium reacted and was released in the liquid phase even at the initial sprayed days.

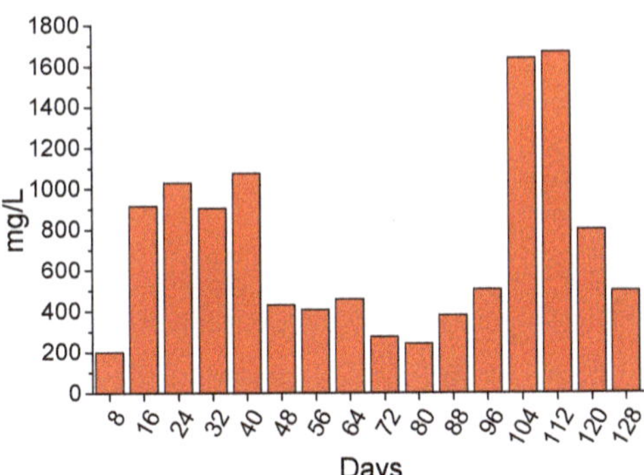

Figure 9. The evolution of magnesium content (mg/L) of rinsing liquids at 8-days intervals during the application of long duration acid spraying test.

According to the above-mentioned results, the long duration test showed that the examined coating was consumed during the experimental procedure. The by-products can be evaluated by the following XRD results.

3.4. XRD Analysis

3.4.1. 4-Days Accelerated Acid Spraying Tests

The respective samples were examined by using XRD and the respective results are presented as XRD diffractograms overlay. Figure 10 shows the results regarding the coated specimens after 4 days of each acid spraying test, i.e., the HHD test, the chamber test using coated concrete specimens and the chamber test using PMMA plates as substrate. Moreover, the diffractogram of the raw material, i.e., without being subjected to acid spraying, is also presented in the same Figure for comparison reasons.

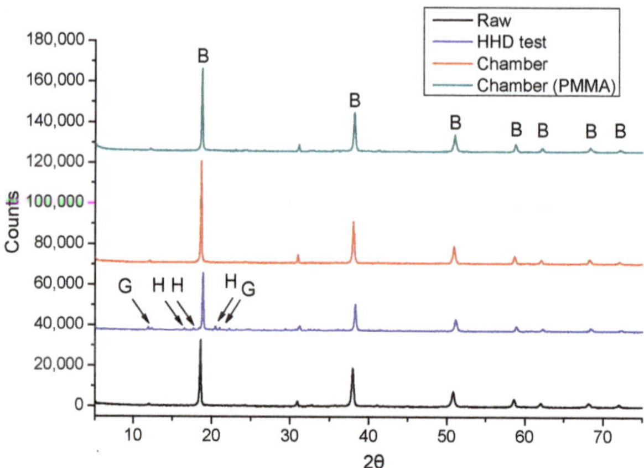

Figure 10. XRD overlay diffractograms of coated specimens after 4 days of different acid spraying tests' application (the raw material is also presented); B: Brucite, G: Gypsum, H: Hexahydrite.

According to these results, the main peaks of brucite ($Mg(OH)_2$) are indicated, whereas the peaks of other phases (i.e., lizardite ($Mg_3(Si_2O_5)(OH)_4$) and dolomite ($Ca,Mg(CO_3)_2$)) are not easily indicated (presenting low intensity), but they are also included in the semi-quantitative results presented in Figure 11. The main corrosion by-products, i.e., gypsum ($CaSO_4 \cdot 2H_2O$) and hexahydrite ($MgSO_4(H_2O)_6$), are indicated in Figure 10, but only for the case of HHD test. This fact could be attributed to the different frequency of applied spraying and the higher acid concentration, as these were the main differences between the applied tests. The concentration of sulfuric acid during the HHD test was higher (4 M), hence it may lead to the immediate formation of by-products, as comparing to the other tests' samples, where much lower sulfuric acid concentration was used; noting also that the real conditions, existing in sewer pipe ("crown") surface, are much milder, than the high concentrations used during the HHD test. The lower applied concentration of acid solution during the chamber tests, allowed the faster reaction of the coating with the acid; hence preventing the diffusion of acid through the coating. As a result, the concrete surface was not affected, and no corrosion products were formed in the later cases, i.e., during the chamber tests.

It is important to note that the formed magnesium sulfate (hexahydrite) is water soluble and can be easily washed out, due to spraying applications; hence, it cannot always be traced. The differences between the tested specimens, when compared to the raw (initial) material used, indicate the formation of by-products, specifically in the case of HHD test sample. Additionally, the substrate did not seem to interfere with the found crystallographic phases results, as no peaks of concrete phases were detected, such as calcite or quartz.

The semi-quantified results of XRD diffractograms (Figure 11) confirm the presence of main by-products (gypsum and hexahydrite) in the HHD test samples. It can be also observed that brucite was in excess quantities in all accelerated tests, indicating that the coating could offer protection to the substrate, even after the end of 4-day spraying test procedures.

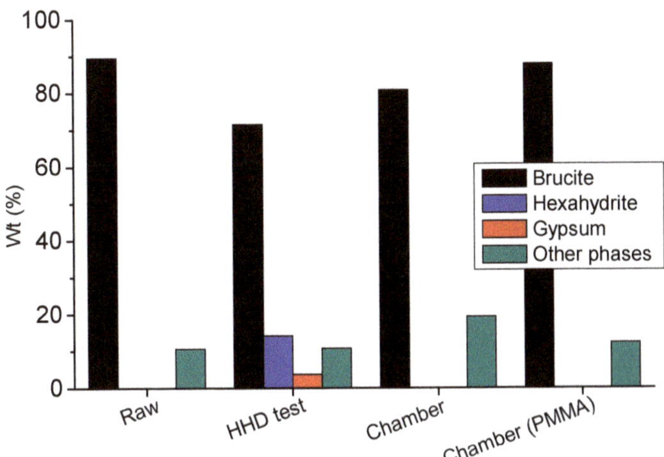

Figure 11. The semi-quantitative results of XRD analysis, regarding the presence of crystalline phases for the coated specimens after 4 days of each acid spraying test applications; «raw» is the coating but without being subjected to spraying applications.

Due to the two different sources of calcium, i.e., from the concrete or from the coating, it is not clear whether gypsum was a corrosion by-product of the substrate (concrete), or of the coating content only. In that way, the tests using PMMA plates were also performed to verify the origin of potential gypsum formation. Regarding the respective results of PMMA plates, gypsum formation was not observed, concluding that the formed gypsum during the HHD tests corresponded to the corrosion of concrete. In comparison, the results of the chamber tests indicate that the coating can provide anti-corrosion protection to the concrete substrate, because gypsum formation was better prevented in this case.

3.4.2. 4-Months Accelerated Acid Spraying Test

The coating was examined under sulfuric acid spraying in a spraying chamber test for 4 months, using 0.1 M sulfuric acid solution. The selected XRD diffractograms at days 0, 32, 64, 96, and 128 are presented in Figure 12. The peaks of brucite were identified, along with the gypsum for the samples at days 96 and 128. No hexahydrite peaks were identified in any sample. Additionally, the peaks of the other phases are rather small to be indicated, but they were included in the respective semi-quantitative results presented in Figure 13.

The semi-quantitative XRD results of all eight-day intervals are presented in Figure 13. The crystalline phase of brucite (coating) was found in excess until the sample of day 120. This result indicates that the coating can protect the substrate almost for the time duration that it was theoretically calculated; noting that the total amount of sprayed acid was calculated in order to consume the whole coating mass. Therefore, it can be expected that gypsum would be formed, due to the consumption of protective coating and the subsequent reaction with the exposed concrete surface before the end of this test.

The gypsum crystalline phase was identified at the sample of day 64, but at low content (i.e., close to experimental error), although the periodic increase of gypsum was recorded even after the 88th day. The gypsum at day 64 might be formed due to sample weaknesses and structure deformation, creating defects and leading to the detachment of coating.

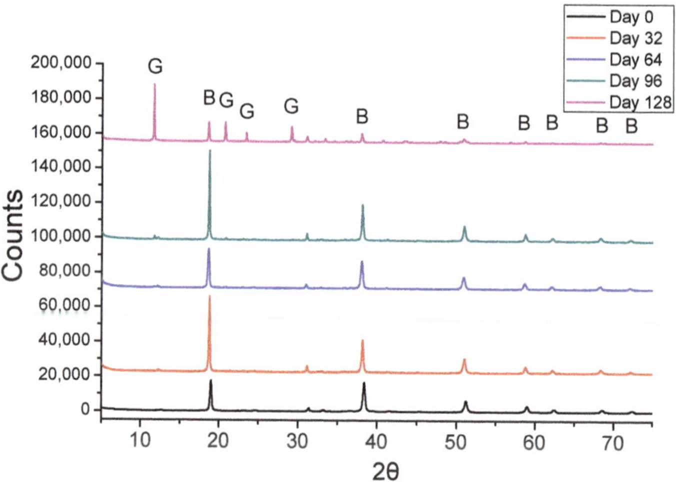

Figure 12. XRD overlay diffractograms of selected coated concrete specimens during the application of the 4-months acid spraying test; G: Gypsum, B: Brucite.

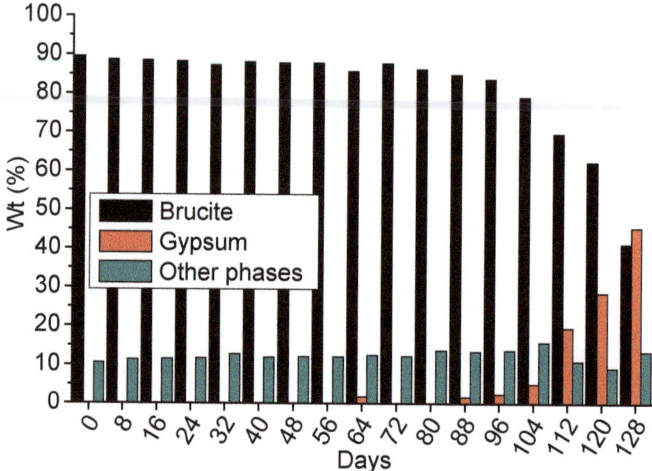

Figure 13. The semi-quantitative results of XRD analysis, regarding the crystalline phases of coated concrete specimens during the application of 4-months acid spraying test.

According to these results, the coating can practically offer long-term anti-corrosion protection to the concrete surface. The coating protected the concrete surface for at least the 87.5% of the calculated theoretical time without allowing the formation of significant gypsum by-product. After day 112, the degradation of concrete is recorded, due to gypsum formation, despite the maintained alkaline pH values.

3.5. Attenuated Total Reflectance

The ATR measurements record the vibrations of bonds of the compounds and the respective spectra are displayed in Figure 14. The representative spectra were selected to reveal the progress of gypsum formation. In particular, the spectra of samples at 8, 64, 80, and 128 days are presented. The stretching of hydroxylic groups of magnesium hydroxide bonds (brucite) is shown at the wavenumber 3694 cm^{-1} [35]. The peaks, due to carbonate ion, appear at 1459 cm^{-1} and 879 cm^{-1}, which implies that carbonated species are also present in the samples, although the relevant IR response is rather weak.

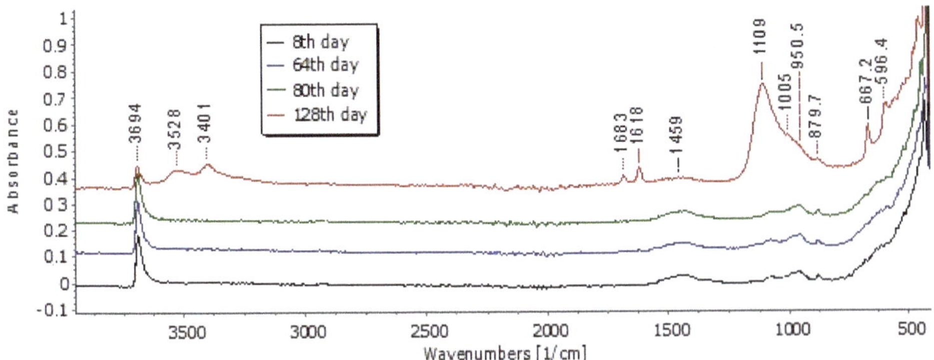

Figure 14. The evolution of ATR spectra of selected samples at 8, 64, 80 and 128 days during the long-term spraying test.

At the 128th day spectrum gypsum peaks were identified. The strong peak at 1109 cm^{-1} is attributed to sulphates (SO$_4$$^{-2}$) [36]. Additionally, the band at 1683 cm^{-1} and 1618 cm^{-1} is attributed to the O-H bending vibrations of water molecules [35,37]. The sharp peaks at 667 cm^{-1} and 596.4 cm^{-1} are owed to the stretching and bending modes of sulfate group of the gypsum [37]. Gypsum is formed and recorded at later dates of the long-term testing, showing that the ATR measurements are in good agreement with the aforementioned XRD results.

4. Discussion

According to the previous experimental study [28], the characteristics of magnesium hydroxide-based coatings for the protection of concrete surfaces can determine its behavior and effectiveness. The particle size, purity, and specific surface area of raw material (i.e., the initial magnesium oxide powder) and the additives used for the slurry preparation can influence the performance of the coating. The studied magnesium hydroxide was designed to produce a coating that can offer better anti-corrosion protection in comparison with other relevant materials, when applied on the (internal) surface of concrete sewer pipes. The PSD and the purity of used hydroxide were selected, according to the results of the relevant previous study, while the SSA was selected to be low, in order to achieve the desired (low) viscosity of the slurry, as this parameter did not seem to affect particularly the resulting anti-corrosion properties of coating [28].

The particle size of magnesium hydroxide was selected to be relatively small in order to enforce the interacting ability of resulting coating, based on the results of the previous relevant research [28]. Although the studied hydroxide presents a smaller particle size than the previous samples, the presence of hexahydrite was found to be in a smaller amount than expected, due to washing out. Hand-held testing resulted in hexahydrite and gypsum formation on the surface of coated specimens as a result of higher acid concentration that was directly sprayed on this surface (Figure 11).

Another crucial property that was selected according to previous evaluation [28] was the purity of used raw material. The magnesium hydroxide content in this case was higher, and it added to the better preservation of alkaline surface values. According to pH measurements, the values were maintained above pH 8.0 in all examined cases.

The slurry was applied on concrete specimens for long- and short-time evaluation of its resistance during acid spraying, as well as on PMMA specimens for the study of coating reaction (neutralization with sulfuric acid), but without the interference of concrete substrate.

The resulting surface pH of studied coating was maintained in the alkaline region (i.e., pH > 8.0), regardless the duration of testing and the concentration of sprayed acid. Furthermore, the coating maintained the pH to the alkaline region, when tested on both PMMA and concrete substrates and for long period of time (128 days), by applying lower acid concentration, as well as for shorter period of time (4 days), by applying higher acid concentration. The latter indicates the resistance of coating to neutralization for the aforementioned different experimental conditions.

The adhesion between the coating and the concrete surface seems to result in a system, where the two materials can "co-operate" well and the coating protects efficiently and effectively the concrete substrate. The respective mass change results pictured a gradual removing of coated material and confirmed that most probably hexahydrite was washed out, as the mass of coatings was decreased in both cases, i.e., for the concrete and PMMA substrates.

The long-term testing revealed the behavior of coating under approximately real conditions, existing in sewer pipes. The simulation of lower acid concentration (0.1 M) showed that efficient protection can be achieved. The gypsum formation was prevented, according to the XRD and ATR results. More specifically, gypsum was identified by XRD at the 112th day of spraying and ATR verified its late formation. Therefore, the coating can protect completely the concrete for this time period, under the applied experimental conditions. After 112 days the gypsum formation started, accompanied with extended brucite consumption, which indicates that the coating's degradation leads to gypsum formation, i.e., strong evidence that the concrete is no longer protected. The magnesium removal from the surface and its leaching in washouts seemed to follow a nonlinear behavior, as it was found quite intensive during the initial experimental procedure (i.e., 10–40 days) and later (i.e., after 104 days), according to the quantification of Mg content.

The comparison of different substrates coated with the examined material revealed the gypsum source. The XRD results of PMMA coated plates did not show any gypsum formation. As a result, the gypsum formation does not originate from the calcium content of coating material, but it is attributed to the corrosion of concrete surface, due to the presence (spraying) of sulfuric acid. Additionally, it was proven that the coating's adhesion onto the concrete's surface also plays an important role on the consumption of coating.

5. Conclusions

The magnesium hydroxide slurry was found to protect the concrete from MIC by effectively maintaining alkaline surface pH values and reacting with (neutralizing) the sprayed sulfuric acid. The results of the present study indicated that the medium particle size of the raw magnesium hydroxide material used is optimal for the studied experimental conditions, in order to achieve adequate interacting ability with the sprayed sulfuric acid. Improved purity can also enhance the alkaline and anti-corrosion properties of produced coating. Moreover, the SSA property seemed to be rather independent of the anti-corrosion behavior of coating.

Corrosion by-product (i.e., gypsum) was only formed after spraying with higher concentration of sulfuric acid, during the hand-held device accelerated acid spraying test (i.e., 4 M). However, during the accelerated tests using lower sulfuric acid concentration (i.e., closer to the conditions existing in a real sewer pipe), no production of gypsum was observed. The long duration (accelerated) acid spraying test indicated that the respective coating can protect the concrete substrate, by reacting (consuming) with sulfuric acid, under conditions that can simulate the sewer pipe environment.

To sum up, the improved properties of magnesium hydroxide raw material can enhance the anti-corrosion behavior of produced coating and preserve effectively the properties of concrete substrate, even during long-term test procedures, and as a result also when used in actual sewer pipes.

Author Contributions: Conceptualization, E.A. and A.Z.; methodology, D.M. and E.-C.T.; validation, D.M. and E.-C.T.; investigation, D.M. and E.-C.T.; resources, E.A. and A.Z.; data curation, D.M. and E.-C.T.; writing—original draft preparation, D.M. and E.-C.T.; writing—review and editing, E.A. and A.Z.; supervision, E.A. and A.Z.; project administration, E.A. and A.Z. All authors have read and agreed to the published version of the manuscript.

Funding: This research has been co-financed by the European Union and Greek national funds through the Operational Program Competitiveness, Entrepreneurship and Innovation, under the call RESEARCH–CREATE–INNOVATE (project code: T1EDK-02355).

Co-financed by Greece and the European Union

Institutional Review Board Statement: Not applicable.

Informed Consent Statement: Not applicable.

Data Availability Statement: The data underlying this article will be shared on reasonable request from the corresponding author.

Acknowledgments: The authors wish to thank the Grecian Magnesite S.A. Company for the supply of raw materials.

Conflicts of Interest: The authors declare no conflict of interest.

References

1. Olmsted, F.H.; Hamlin, H. Converting Portions of the Los Angeles Outfall Sewer into a Septic Tank. *Eng. News* **1900**, *44*, 317–318.
2. Parker, C. The corrosion of concrete. The isolation of a species of bacterium associated with the corrosion of concrete exposed to atmospheres containing hydrogen sulfide. *Aust. J. Exp. Biol. Med. Sci.* **1945**, *23*, 81–90. [CrossRef]
3. Parker, C. The corrosion of concrete. The function of Thiobacillus Concretivorus in the corrosion of concrete exposed to atmospheres containing hydrogen sulfirde. *Aust. J. Exp. Biol. Med. Sci.* **1945**, *23*, 91–98. [CrossRef]
4. Thistlethwayte, D.K.B. The control of sulphides in sewerage systems. *AIChE J.* **1972**, *19*, 173. [CrossRef]
5. Pomeroy, R.D. *Process Design Manual for Sulfide Control in Sanitary Sewerage Systems*; US Environmental Protection Agency, Technology Transfer: Washington, DC, USA, 1974.
6. Wu, M.; Wang, T.; Wu, K.; Kan, L. Microbiologically induced corrosion of concrete in sewer structures: A review of the mechanisms and phenomena. *Constr. Build. Mater.* **2020**, *239*, 117813. [CrossRef]
7. Wu, W.; Zhang, F.; Li, Y.; Song, L.; Jiang, D.; Zeng, R.C.; Tjong, S.C.; Chen, D.C. Corrosion resistance of dodecanethiol-modified magnesium hydroxide coating on AZ31 magnesium alloy. *Appl. Phys. A Mater. Sci. Process.* **2020**, *126*, 8. [CrossRef]
8. Hvitved-Jacobsen, T.; Vollertsen, J.; Nielsen, A.H. *Sewer Processes: Microbial and Chemical Process Engineering of Sewer Networks*; CRC Press: Boca Raton, FL, USA, 2013; ISBN 9781439881774.
9. Okabe, S.; Odagiri, M.; Ito, T.; Satoh, H. Succession of Sulfur-Oxidizing Bacteria in the Microbial Community on Corroding Concrete in Sewer Systems † Downloaded from. *Appl. Environ. Microbiol.* **2007**, *73*, 971–980. [CrossRef]
10. Noeiaghaei, T.; Mukherjee, A.; Dhami, N.; Chae, S.-R. Biogenic deterioration of concrete and its mitigation technologies. *Constr. Build. Mater.* **2017**, *149*, 575–586. [CrossRef]
11. Li, W.; Zheng, T.; Ma, Y.; Liu, J. Current status and future prospects of sewer biofilms: Their structure, influencing factors, and substance transformations. *Sci. Total Environ.* **2019**, *695*, 133815. [CrossRef]
12. Wells, T.; Melchers, R.E. *Findings of a 4 Year Study of Concrete Sewer Pipe Corrosion*; Australasian Corrosion Association: Victoria, Australia, 2014.
13. Wells, T.; Melchers, R.E.; Bond, P. *Factors Involved in the Long Term Corrosion of Concrete Sewers*; Australasian Corrosion Association: Victoria, Australia, 2009; Volume 11, pp. 345–356.
14. Zhang, G.; Xie, Q.; Ma, C.; Zhang, G. Permeable epoxy coating with reactive solvent for anticorrosion of concrete. *Prog. Org. Coat.* **2018**, *117*, 29–34. [CrossRef]
15. Aguiar, J.B.; Camões, A.; Moreira, P.M. Coatings for Concrete Protection against Aggressive Environments. *J. Adv. Concr. Technol.* **2008**, *6*, 243–250. [CrossRef]
16. Roghanian, N.; Banthia, N. Development of a sustainable coating and repair material to prevent bio-corrosion in concrete sewer and waste-water pipes. *Cem. Concr. Compos.* **2019**, *100*, 99–107. [CrossRef]
17. Khan, H.A.; Castel, A.; Khan, M.S.H. Corrosion investigation of fly ash based geopolymer mortar in natural sewer environment and sulphuric acid solution. *Corros. Sci.* **2020**, *168*. [CrossRef]

18. Herisson, J.; Guéguen-Minerbe, M.; van Hullebusch, E.D.; Chaussadent, T. Influence of the binder on the behaviour of mortars exposed to H2S in sewer networks: A long-term durability study. *Mater. Struct.* **2017**, *50*, 8. [CrossRef]

19. Ganigue, R.; Gutierrez, O.; Rootsey, R.; Yuan, Z. Chemical dosing for sulfide control in Australia: An industry survey. *Water Res.* **2011**, *45*, 6564–6574. [CrossRef] [PubMed]

20. Liu, Y.; Sharma, K.R.; Ni, B.J.; Fan, L.; Murthy, S.; Tyson, G.Q.; Yuan, Z. Effects of nitrate dosing on sulfidogenic and methanogenic activities in sewer sediment. *Water Res.* **2015**, *74*, 155–165. [CrossRef]

21. Wang, T.; Wu, K.; Kan, L.; Wu, M. Current understanding on microbiologically induced corrosion of concrete in sewer structures: A review of the evaluation methods and mitigation measures. *Constr. Build. Mater.* **2020**, *247*, 118539. [CrossRef]

22. Almusallam, A.A.; Khan, F.M.; Dulaijan, S.U.; Al-Amoudi, O.S.B. Effectiveness of surface coatings in improving concrete durability. *Cem. Concr. Compos.* **2003**, *25*, 473–481. [CrossRef]

23. Berndt, M.L. Evaluation of coatings, mortars and mix design for protection of concrete against sulphur oxidising bacteria. *Constr. Build. Mater.* **2011**, *25*, 3893–3902. [CrossRef]

24. Duszczyk, J.; Siuzdak, K.; Klimczuk, T.; Strychalska-Nowak, J.; Zaleska-Medynska, A. Manganese Phosphatizing Coatings: The Effects of Preparation Conditions on Surface Properties. *Materials* **2018**, *11*, 2585. [CrossRef]

25. Aguirre-Guerrero, A.M.; Mejía de Gutiérrez, R. Alkali-activated protective coatings for reinforced concrete exposed to chlorides. *Constr. Build. Mater.* **2021**, *268*, 121098. [CrossRef]

26. Sydney, R.; Esfandi, E.; Surapaneni, S. Control Concrete Sewer Corrosion via the Crown Spray Process. *Water Environ. Res.* **1996**, *68*, 338–347. [CrossRef]

27. James, J. Controlling sewer crown corrosion using the crown spray process with magnesium hydroxide. *Proc. Water Environ. Fed.* **2003**, *2003*, 259–268. [CrossRef]

28. Merachtsaki, D.; Tsardaka, E.-C.; Anastasiou, E.K.; Yiannoulakis, H.; Zouboulis, A. Comparison of Different Magnesium Hydroxide Coatings Applied on Concrete Substrates (Sewer Pipes) for Protection against Bio-Corrosion. *Water* **2021**, *13*, 1227. [CrossRef]

29. European Committee for Standardization. *EN 1766: 2017 Products and Systems for the Protection and Repair of Concrete Structures—Test Methods—Reference*; European Committee for Standardization: Brussels, Belgium, 2017.

30. European Committee for Standardization. *EN 1916: 2002: Concrete Pipes and Fittings, Unreinforced, Steel Fibre and Reinforced*; European Committee for Standardization: Brussels, Belgium, 2002.

31. Diamanti, M.V.; Brenna, A.; Bolzoni, F.; Berra, M.; Pastore, T.; Ormellese, M. Effect of polymer modified cementitious coatings on water and chloride permeability in concrete. *Constr. Build. Mater.* **2013**, *49*, 720–728. [CrossRef]

32. Merachtsaki, D.; Fytianos, G.; Papastergiadis, E.; Samaras, P.; Yiannoulakis, H.; Zouboulis, A. Properties and Performance of Novel Mg(OH)2-Based Coatings for Corrosion Mitigation in Concrete Sewer Pipes. *Materials* **2020**, *13*, 5291. [CrossRef] [PubMed]

33. European Committee for Standardization. *EN 1542: 1999: Products and Systems for the Protection and Repair of Concrete Structures—Test Methods—Measurement of Bond Strength by Pull-Off*; European Committee for Standardization: Brussels, Belgium, 1999.

34. European Committee for Standarization. *EN 13578: 2003: Products and Systems for the Protection and Repair of Concrete Structures—Test Methods—Compatibility on Wet Concrete*; European Committee for Standardization: Brussels, Belgium, 2003.

35. Fockaert, L.I.; Würger, T.; Unbehau, R.; Boelen, B.; Meißner, R.H.; Lamaka, S.V.; Zheludkevich, M.L.; Terryn, H.; Mol, J.M.C. ATR-FTIR in Kretschmann configuration integrated with electrochemical cell as in situ interfacial sensitive tool to study corrosion inhibitors for magnesium substrates. *Electrochim. Acta* **2020**, *345*, 136166. [CrossRef]

36. Nasrazadani, S.; Eghtesad, R.; Sudoi, E.; Vupputuri, S.; Ramsey, J.D.; Ley, M.T. Application of Fourier transform infrared spectroscopy to study concrete degradation induced by biogenic sulfuric acid. *Mater. Struct.* **2016**, *49*, 2025–2034. [CrossRef]

37. Ashrit, S.; Chatti, R.V.; Nair, U.G. An infrared and Raman spectroscopic study of yellow gypsum synthesized from LD slag fines. *Mater. Sci. Eng.* **2018**, *2*. [CrossRef]

MDPI
St. Alban-Anlage 66
4052 Basel
Switzerland
Tel. +41 61 683 77 34
Fax +41 61 302 89 18
www.mdpi.com

Materials Editorial Office
E-mail: materials@mdpi.com
www.mdpi.com/journal/materials

www.ingramcontent.com/pod-product-compliance
Lightning Source LLC
LaVergne TN
LVHW070855170726
843515LV00005B/654